All the best wishes

Kotton, Port, Rail Center: A History of Early Radio in Houston

By

Christopher Varela

Fez Publishing

Houston, Texas

This book is printed on acid free paper.

Published by Fez Publishing, Houston, Texas.

Library of Congress Number: 2004091674

ISBN: 0-9748060-0-5

Design and Layout by Budget Book Design

Printed by Books Just Books

ACKNOWLEDGEMENTS

Though this was a self-published project, it was by means a solo effort. I would like to thank those individuals (some of who are no longer with us) who shared their personal histories on the subject. Their wonderful stories are what fueled my desire to complete the project and make it known.

My deepest gratitude goes to Larry Lamar Lawson (KK5KE) who initially introduced me to the radio broadcasting history of Houston. From the conversations we had, I knew that a work needed to be done on the subject.

Thanks to Anita Darras who showed me the functions of Microsoft Word and how to format my work into a book.

Finally, I would like to acknowledge my wife, Imelda, who supported me and had the patience to see me through this effort.

I dedicate this book to my sons, family and friends with the hopes that they follow and achieve their dreams.

Introduction

If one is to criticize Houston's lack of appreciation for its twentieth century history, then perhaps the city's most underappreciated and least understood aspect is its early radio broadcasting scene. This era, filled with advancements in wireless communications, forever changed the way we would share and assimilate information. The technology behind radio has paved the way for all things related to transmitting radio signals including remote controlled devices, cellular telephone advancements, satellite communications and television.

Radio has left its unique mark on Houston that can still be seen today. Some of Houston's most famous luminaries from the twentieth century involved themselves with radio and eventually broadcasting. They and other lesser-known Houstonians introduced radio communication and helped it evolve into a social phenomenon that entertains and informs audiences scattered in remote areas. From this generation of radio pioneers came the licensing of call letters for three early broadcasting stations: KPRC, KTRH and KXYZ, stations that still thrive in Houston's market today.

Yet no memorial stands for many of these unsung individuals. Even more unfortunate is that the history behind these visionary men and efforts to promote radio has been sparsely documented with information that is often incorrect, until now. *Kotton, Port, Rail Center: A History of Early Radio in Houston* explores the city's vibrant radio scene before the Great Depression. This history will cover radio technology at its infancy and how it evolved into the entertainment and information forum set against a Houston backdrop of the early twentieth century.

Several years ago, a former coworker and I had a discussion concerning the first radio broadcasting stations in Houston that prompted me to investigate the subject more deeply. Radio, more than any other medium fascinates me. More than television, radio provides a personal nostalgia as I remember often spending hours of sleepless nights in bed listening to my portable radio entertain with disembodied "voices" and music. Even today, I can appreciate more the drama of a trivial weather report from a radio broadcast than one aired from a newscast on the television.

After initial research into Houston's past and its connection to radio, I learned of an engaging history forgotten and for the most part, undocumented. A book has been written about Houston's early television industry, so why not one about the city's early radio broadcasting era? Without radio, television, at the very least, would not have been accepted as quickly as it was. Radio was a necessary predecessor to television. It laid the technological foundation for how television signals would be transmitted and it created the "broadcast format" that would be taken a step further with television.

On a social level, radio broadcasting brought about some profound ramifications. Music that usually was heard only by certain Houston ethnic and social groups was made more available to more diverse and larger groups. Entertainment that was often only available to affluent people could now being accessed by middle class Houstonians as they sat at home with their radio receivers. Those less fortunate to own a radio receiver at their residence could still experience broadcasting by attending social events, often gathered around a department store's radio set display or the ever popular radio party held at a neighbor's home. Houstonians were becoming better informed of current news events and situations and, more often, in real time. Radio was the first modern invention to make our world a little smaller.

Since Houston's radio history is complex, this history covers only the time periods when radio communication was introduced to the city and how radio expanded to become a broadcasting phenomenon before 1930. To cover all of Houston's radio history of the twentieth century would require more that one written work.

Contrary to what has been recorded by historians, Texas did not embrace radio technology in 1911 but actually as early as 1904. During this era, controversial scientists and unethical businessmen sought to exploit radio communication and Americans' ignorance of it by installing, often faulty, radio facilities in as many places as possible mostly for quick, financial gain and less for the advancement of science. Texas' first fully functional radio facilities were in Houston and two other upper Gulf coast cities, whereby Texans were now being introduced to a new form of long distance communication without the use of telegraph and telephone lines. Other reputable groups in Texas' young oil industry experimented with radio communication at this time too, with more successful results.

Following the inevitable fate of these unscrupulous wireless companies, the emerging party on the scene was an informal network of young men and boys with the common interest of communicating with others on radio. Known as amateur operators, these middle-class youth sought drama and adventure in a still mostly unregulated radio realm. Often rivaling the expertise and technology of professional radio industries of the day, the amateur operator dominated radio communication in the 1910s and restored the merit of their science. They advanced their passion and art to a new level. Returning as heroes from World War I, the amateur operator invented radio broadcasting initially for amusement, not realizing that what they were doing was creating a new social phenomenon.

By 1921, the regular broadcasting and listening of music and informational briefs was common practice for the amateur operator yet still a relatively unknown phenomenon to the average lay person. In just the next year though, radio broadcasting became a full-blown social event including groups never before involved with radio now establishing broadcasting stations.

The first broadcasting stations in Houston started in an automobile service garage, a middle-class household in today's midtown, and a palatial movie theater in downtown. These pioneering stations provided the Houston listener with a day full of entertainment ranging from live music and dance parties until midnight to Christian programming, local sporting events and experts discussing health and women's issues.

In just two years, broadcasting transformed from an eccentric act performed by a few to a high-priced event using the latest transmitting equipment for advertising gains. Soon, big Houston enterprises such as newspapers and luxurious hotels sought to install broadcasting stations for their advertising potential. National companies such as the National Broadcasting Company (NBC) and the Columbia Broadcasting Station (CBS) were formed in the late 1920s dedicated to bringing radio programming to Americans around the country with a new technique of "chaining" stations through long distance communication lines. By the late 1920s, two stations in Houston affiliated themselves with these networks to provide Houstonians with a flavor of national broadcasts.

Much of the programming and radio personalities we commonly credit to the golden age of radio during the 1930s actually aired before 1929. For example, in Houston, jazz music found an unexpected advocate in broadcasting. More diverse groups through the advent of broadcasting were now hearing this new form of American music. In just months, broadcasting changed a relatively unknown amateur operator into a household name heard around town as he served as a radio announcer. Cub sports writers for local newspapers became on-air sports analysts giving the play-by-play as it happened. Vaudeville thespians that for years had entertained Houston audiences were drawn to broadcasting and became some of history's first radio personalities, bringing frolic and comedy to broadcasting.

An array of characters including scientists, young men and boys, mavericks, World War I heroes, legendary reclusive billionaires, ambitious philanthropists, Texas governors, and other colorful personalities participated in this complex history.

Certain terminology used in this book needs to be clarified for the contemporary reader. Initially, "radio" was also referred to as "wireless." This book will interchange the terms "radio telegraphy" and "wireless telegraphy" to refer to the exchange of messages between two parties heard over the airwaves as dots and dashes and communicated on the international system known as Morse code. In time, the ability to transmit speech and music slowly supplanted telegraphy and the means to radio messages. This technology will be referred to as "radio telephony" or "wireless telephony." The term "broadcast" did not come on the scene until about 1922. Before then, the transmission of music was coined "wireless concert," "radio concert" or "radiocast."

All wireless communications and transmissions covered here were conducted on the AM radio band. Frequencies for stations before the mid-1920s were measured on the AM band in units known as "meters." Not exactly an accurate measurement, "meters" were numerical units usually divisible by ten. As more broadcasting stations signed on during the 1920s, a more precise measurement called the "kilocycle" (abbreviated as kc) had to be used to designate a station's frequency. Later, an international agreement was adopted to replace "kilocycle" with the term we use today to measure AM frequencies, "kilohertz" (abbreviated as kHz).

Chapter One

Sixteen-year-old Alfred Patrick Daniel rode his bicycle to a towering structure that rose two hundred and ten feet into the air. He had spotted the structure from his tree house about a mile away and decided to have a closer look. The year was 1906[1] in Houston, Texas and the hot, humid days of summer lingered as the season was drawing to a close in August.

As he approached the structure, Alfred Daniel may have already known that what he had seen from a bird's eye view at his residence was a radio tower being built for a wireless station of the American De Forest Wireless Telegraph Company.[2] In the last few months, local newspapers had steadily covered this wireless company coming to town as it attempted to amend one of its disappointing ventures two years before in Texas. Daniel may have also understood the concept of radio or wireless telegraphy, as it was more commonly known as then. For among his contemporaries, an interest in harnessing and manipulating radio waves for a new form of long distance communication was quickly gaining serious interest among certain business circles. One thing is certain though, after that day, the course for Houston's early days in radio broadcasting was set into motion when this teenager arrived on a bike to learn about wireless.

At the turn of the twentieth century, wireless communication was in its infancy. During the 1800s, a few scientists discovered and theorized the ability to send intelligible signals through the free atmosphere between two points without the use of wire connections. Some obscure, colorful characters took it a step further, turning the theoretical into reality. But these incidents were isolated with no long-term implications. Then a German scientist named Heinrich Hertz demonstrated the practical ability to generate and receive radio waves in the late 1880s. Eventually, Hertz's study of radio waves led to the discovery that they traveled at incredible speeds and were able to pass through certain solid objects. Radio waves became known as Hertzian waves. Soon, others perfected Hertz's equipment and made the use of radio waves practical.

Enter Guglielmo Marconi: a young, Italian-Irish scientist who studied these nineteenth-century innovators and perceived the marketability of wireless. After successfully demonstrating his wireless system in Europe, Marconi was invited to America in 1899 where he reported that year's America Cup yacht races from a radio transmitter on a yacht to a receiving land station. In 1901, he would again grab headlines as he was the first to successfully radio transmit the letter *S* across the Atlantic Ocean. Hailed as a scientific hero, Marconi seized the opportunity to launch an incorporated venture, a business dedicated to selling his long distance communication to anyone appreciating the viability of his idea.

Competitors, of course, rose to the occasion to capitalize on Marconi's dream. One of these was Lee de Forest. A graduate of the Sheffield Scientific School of Yale University, de Forest sought to create a profitable enterprise from his years of wireless experimentation. Unfortunately, he fell in with the wrong business people. While Marconi had his business established, de Forest joined with Wall Street broker Abraham White and incorporated the American De Forest Wireless Telegraph Company in 1902.[3] Described as a flamboyant figure, Abraham White was already known as a disreputable businessman when he joined de Forest. White, a native Texan from the town of Corsicana,[4] was originally named Schwartz before he entered the business world as a corrupt stockbroker. De Forest's autobiography also points out that White added a more colorful connotation to his name by calling himself Abraham "Lincoln" White.[5] Furthermore, his peers nicknamed him "Postage Stamp"

White, stemming from a Wall Street scheme that netted him a fortune by selling U.S. government bonds with a minute investment outlay of postage costs.[6]

From the beginning, White had no intentions of furthering the advancement of wireless for the benefit of science. And de Forest fully understood this, so long as the financial backing was in place to continue with his research and development. Unfortunately, de Forest quickly contracted the greed that consumed White. They saw the relatively untapped commodity of radio communication as "bait" to lure naïve investors into their "get rich quick" scheme. Their idea was to sell as much stock as possible of the American De Forest Wireless Telegraph Company with little concern for the advancement of wireless. Thus, they dedicated little resources to research and development and the majority of the money went into their pockets (though mostly into White's).

To promote stock sales, they staged grand wireless demonstrations around highly visible places. They also mailed stock promotional material to potential investors. Around the East Coast and the Midwest, the American De Forest Wireless Telegraph Company began installing large-scale wireless stations. In 1904, as part of a scheme to attract investors, a wireless apparatus was installed in an automobile parked outside of Wall Street where stock quotes were received by radio transmissions. Meanwhile, the United States Navy took an interest in the company's activities and initially contracted with de Forest to install radio equipment in some of the fleet. Stock sales for the company were brisk.

However, underneath the successful marketing propaganda and the promise of wireless was a thinly veiled array of problems. Many of the company's stations transmitted wireless messages poorly and others never even sent a single message. One station that was reported to have cost $3,000 to build brought in $50,000 in stock sales and never transmitted any messages.[7] And, as expected, when one con man's avarice and greed exceeds that of the other, tensions began to build in the

White-de Forest partnership. White continued skimming money from the company and was not accurately reporting the "profits" to de Forest who was gradually being forced out by his partner.

Despite his greed, Lee de Forest was a wireless inventor first. Originally, it was his intention to work in research and development and leave the business activity to White. Indeed, as scientific director, de Forest supervised what progress the company made in radio technology and created a large number of stations in many parts of the nation. One thing he was eager to prove was that radio could be used as a successful long distance, overland wireless system,[8] a feat considered at the time by some to be a difficult, if not impossible task. During the height of the American De Forest Wireless Telegraph Company, a 300-foot high radio tower with de Forest's name in large letters on its side was installed in St. Louis for that city's huge 1904 World Fair Exposition, a centennial celebration of the Louisiana Purchase.[9]

From St. Louis, de Forest hoped to establish a long distance overland system. Relay stations would be placed through the heart of the Midwest, eventually going through Abraham White's native state of Texas and finally ending in Galveston. Rebuilding from the recent destructive hurricane of 1900, Galveston was on the road to recovery and was ready to show the world its resilience. A Galveston exhibit at the World's Fair, where de Forest had his impressive wireless station, demonstrated the city's plans to build a monumental, protective seawall against further hurricanes.[10] Galveston officials excitedly anticipated two city triumphs; to elevate their coastal city behind a blockading seawall and an invitation by the De Forest company to participate in a historic overland wireless feat.

Just prior to the opening of the World's Fair in October, the St. Louis station showed great promise, sending an overland message a record 105 miles away. The following week, it transmitted over 300 miles to Chicago. However, these incidents were considered flukes and the real test remained in establishing a consistent inland wireless communication service. Nevertheless, excitement was high in the company and efforts continued to build a wireless system through Texas to the Gulf Coast.

Lee de Forest (seated) and Abraham White (standing) as they pose at their wireless station built for the 1904 St. Louis World's Fair. Here they would announce their intentions to create an inland chain of stations leading into Texas. *Courtesy of History San Jose.*

A preliminary team of technicians from the De Forest Company arrived in Texas in 1904 to do wireless tests. Included in this Texas venture was H. E. "Pop" Athearn,[11] an early wireless operator and close assistant to de Forest. "Pop" Athearn had just returned from a three-month job of transmitting news of the ongoing Russian-Japanese conflict from a ship in the China Sea to a station on mainland China. The *London Times* newspaper had contracted with the De Forest company to report the war's progress and peace negotiations.

Not necessarily interested in making wireless contact with the other relay stations that were breaking long distance records with the huge St. Louis station, the Texas team hoped to entice interest in the Lone Star State and eventually gain enough support to build larger stations here. While the American De Forest Wireless Telegraph Company was capturing headlines at the World's Fair, work was started in Texas. The Dallas-Fort Worth metropolitan area was selected as the initial testing ground for the experiments.[12] A wireless station was placed in Dallas and a second in Fort Worth. With the short distance between the two cities, impressive wireless communication could be monitored easily here. Investors and technicians in the De Forest Company had high expectations that upon successful testing more stations would be erected in Houston and Galveston.

In September of 1904, the Dallas and Fort Worth stations were declared ready for operation. On the seventh day of that month, messages were exchanged between the cities. The quality of the exchange was described as "exceptionally satisfactory."[13] The excitement increased as two days later the De Forest organization successfully transmitted the initial record-breaking message from St. Louis to Springfield, Illinois, a distance of 105 miles. On September 11 the company extended its wireless system to the record-setting distance of 300 miles. But quickly the excitement in Texas turned to disappointment. On September 14, days after the Dallas-Fort Worth connection began transmitting, a report came out stating that the budding Texas system was sending messages choked with too much static interference and were often indecipherable.

De Forest experts blamed bad location and construction, citing that the radio towers were not high enough to clear the atmospheric interference caused by the existing telephone and telegraph lines.[14] Just a week after the stations went on the air and the announcement of their inferior performance was made, orders were given to take them down and place them in a different location away from congested areas. This marked the beginning of a nationwide downward turn for the American De Forest Wireless Telegraph Company.

However, during these days there was never a shortage of people ready to peddle de Forest's wireless technology in Texas. Even after the publicized embarrassing display of wireless service in Dallas and Fort Worth, de Forest representatives nevertheless turned their efforts toward Houston and Galveston and promised that stations would be installed in these two cities by the end of the year. But by late 1904 assurances that Houston would get a wireless station[15] fell through and another two years would go by before the De Forest organization would establish a station on that city's Buffalo Bayou.

Despite the setbacks, interest in wireless continued to grow in Texas. Months following the debacle of the Dallas-Fort Worth experiment and the impressive overland radio message exchanges from the World's Fair in St. Louis, the De Forest organization dedicated itself to installing a wireless station in Texas' most popular coastal city. In February of 1905, construction began for a Galveston station located in that city's busy wharf section. An article soon followed in the *Galveston Daily News* reporting that equipment from the recently dismantled Dallas station was to be used for the Galveston station.[16]

On March 11, equipped with a 205-foot antenna mast, the station successfully went on the air and communicated with a passenger steamship, the *Concho*, out at sea about forty-five miles from Galveston. Other messages were exchanged that day between the seafaring vessel and what was most likely the first operable long distance wireless station installed in Texas.[17]

Simmering problems began to mount for the De Forest company following the fiasco of the Dallas-Fort Worth stations and the minor victory in Galveston. Three years of litigation culminated in early 1906 when another wireless inventor, Reginald Fessenden, successfully sued the American De Forest Wireless Telegraph Company for infringement rights on a wireless device able to detect and receive radio waves that was being used by de Forest without permission.[18]

Adding to the shady reputation of the company was a new vice president, Christopher Wilson, another former resident of Texas and said to be an ex-colonel in the Confederate army during the Civil War. Like Abraham White, he changed his name to the outrageous Christopher "Columbus" Wilson. According to de Forest, Wilson was even more of a dishonest stockbroker than White with a history of being a callous swindler.[19] Nevertheless, de Forest continued to participate in these fraudulent wireless scams. Little did he know that Wilson and White were conspiring to oust him from the company.

White sent de Forest to Europe in 1906 with hopes of establishing a wireless service there. But Guglielmo Marconi, who already had an incorporated venture there, threatened to sue on infringement grounds. Not wanting to be inundated with another lawsuit, the American De Forest Wireless Telegraph Company retreated from its European venture and de Forest returned to the United States only to be notified of a warrant of arrest on him pertaining to the Fessenden infringement case. On advice from his attorney and Abraham White, de Forest fled to Canada in April 1906 until a bond could be posted for his warrant. The absence of de Forest gave White and Wilson the opportunity to launch a second Texas wireless campaign.

Recently hailed as the scientific wonder of the new century, wireless communication was being ravaged by scheming men and scam jobs. Now, the promising future for this novice technology did not seem so bright.

Christopher "Columbus" Wilson came to Austin, Texas in March 1906 to announce the company's expansion into the state. Bold projections were made that a station would be installed in Dallas in time to cover the results of the state's Democratic convention meeting there in September to select a candidate for state governor. Stations in Houston and Austin were also promised to be in operation in time to receive messages from Dallas.[20]

The company anticipated that twenty-three stations would be scattered throughout the state and in working order by the end of the year.[21] De Forest representatives contended that a wireless system across the state would prove that the years of wireless development would come to fruition and it would be prepared to compete with the telephone and the wire telegraph services. Furthermore, it was alleged that the cost of maintaining a wireless system was also said to be at a competitive level with the logic being that they would not have to pay for the right-of-ways or for any maintenance on miles of wired communication.

The fact remained, though, that the company's reputation to put little money into research and development made their system inefficient as a substitute to wired communications. Even Abraham White confided to close associates that to provide good overland service, many more than twenty-three stations would have to be created,[22] planted in shorter distances from each other.

By now, riddled with controversy and problems, the American De Forest Wireless Telegraph Company may have seen its Texas project as one of its last hopes to redeem itself. And perhaps for good reason did De Forest representatives feel this way. Texas seemed like an ideal place to showcase an impressive wireless system. Abraham White and Christopher Wilson had connections in their native state that could provide invaluable possibilities. None of the existing major wireless companies had entered Texas to conduct their business. Much of the Lone Star State, especially along the Gulf Coast and the eastern region, is flat terrain with few mountains and ideal for providing impressive demonstrations of transmitting radio waves. Furthermore, the budding oil industry in Texas grew attracted to the alternative method of long distance communication that radio could provide. With "new money" entering the state in the form of wildcatters and oil companies, opportunities of luring new wealthy investors were plentiful.

The American De Forest Wireless Telegraph Company commenced in the summer of 1906 to install a wireless station just outside of central Houston.[23] Learning from its mistake two years before, the company chose a site distant from the city's crowded traffic of wired communications. Representatives from the company issued press releases in May that a station had been placed in Austin and others were being simultaneously constructed in the areas of Dallas and Waco with the hopes of them being operational by August. The existing Galveston station had been recently upgraded in strength and was ready to handle the traffic of messages anticipated from the wireless inland plants being built.[24] The company purchased real estate in the community of Brunner, an area roughly two miles west from downtown Houston that would not be annexed by the city until 1915.[25] A tower was constructed along the banks of Buffalo Bayou by connecting several wooden poles to a height of between 210 and 250 feet high for the sensitized wire antenna. The completed mast was then set in a block of mortar or cement. To prevent it from swaying in high winds, supporting guy wires were strung from the mast's tip to the ground.

The main office for the Houston station of the American De Forest Wireless Telegraph Company gave its address as being in the city's downtown on Main Street[26] where the job of promoting the business and selling its stock would be conducted. Messages received in person at the main downtown office would simply be wired to the station on the outskirts of town where they would be sent over radio waves.

But how would Houston react to wireless? As the De Forest company was busy constructing a station, competition from other long distance communication means were stiff in a city seemingly always eager to welcome the newest technological breakthroughs, especially in the communications field. As early as 1838, Texas was being introduced to state-of-the-art communication systems when Samuel Morse offered his new invention, the telegraph system, to Sam Houston for what was then the Republic of Texas. However, President Houston failed to show interest in the invention and it would be decades before Texas would be reintroduced to the telegraph machine.[27] A company called the Texas Telegraph opened a telegraph service between Houston and Galveston in January 1860,[28] one of the first in the nation and the first permanent telegraph lines in the state. Houston got the nation's second telephone system in 1880, just two years after Connecticut. A few years later, Houston's telephone system expanded to serve two hundred telephones, including long-distance service to Galveston. Electricity, the essential element for any communications device, had already been going strong in Houston for three decades before 1906. In 1882, New York City and Houston built the country's first two electric power plants.[29] By the 1880s, some of the city's street corners were lit with electric arc streetlights.

On August 14, 1906, the first official messages from the Houston station were sent. Invited to this groundbreaking ceremony was the city's mayor, Horace B. Rice, who transmitted two messages that day. One was addressed to the mayor of Galveston and received by that city's station. The other was aired to the Democrats' gubernatorial candidate, T. H. Stone, at the state convention in Dallas. Tapped out in Morse code and heard over the airwaves as dots and dashes, it read: "My advice to you is not to attempt to stampede the convention and come in as a dark horse for governor."[30]

Later that day other messages were sent and received, mostly to and from the Galveston station. It is not certain whether or not the Dallas location received the Horace Rice message. Houston and Dallas newspapers made no reports that the two stations were exchanging radio messages. But based on the previous reputation of the company, one can reasonably doubt that the Houston messages did reach Dallas.

At this time another Texas station attempted to reach any existing stations in the state. The newspaper, *Austin Statesman*, did acknowledge that on August 14 and for several days before, the recently placed De Forest station placed near Austin attempted unsuccessfully to reach Houston and Galveston.[31] But Houston had enough difficulties just communicating with Galveston sixty miles away. Just a few days after it signed on, Houston suffered a burned-out transformer and was out of commission for a short period. It was not until after the problems at the Houston station surfaced that the company finally hinted to the press that the Dallas station was still not fully functioning as promised.[32] Attention was deflected from the failures in Dallas and Austin when the furthest recorded station reached by Houston at this time was the one installed by the company in the oil-industry seaport of Port Arthur, Texas, eighty miles away when it went on the air during the first week of October 1906.[33]

Even with its substandard wireless equipment, the company still managed to secure a wireless communication service for the Mallory line of steamships docked in Galveston and Port Arthur to transport passengers to areas as far away as New York. The line's fleet, approximately eleven passenger steamships including the *Concho* and the *Denver*, were equipped with wireless apparatus provided by the De Forest company and were consistently heard hundreds of miles out in the Gulf of Mexico by the Texas land stations at this time.[34] It has been reported by historians that the De Forest wireless service provided on the steamship enterprises was rarely used to generate revenues.[35] Indeed, the company made its profits almost entirely from the stock sales generated by the inland station activities.

The Houston station aired most days of the week and with Galveston and Port Arthur formed the state's first long distance wireless circuit by late 1906. On a frequency between 425 and 450 meters (approximately

870 kHz and 670 kHz, respectively), the three stations regularly exchanged messages and were made available to the public. Given that these Texas stations were considered low- to medium-powered in efficiency and a novelty in the state, apparently little commercial communication was conducted or promoted. Most messages sent and received were for demonstrational in nature. Texans wishing to use the wireless system were probably not charged for the service. Nevertheless, the company took seriously the issue of privacy and security in regards to their messages transmitted over the airwaves. Special codes and abbreviations for the stations were designated in Morse code to ensure that outside listeners could not easily decipher wireless messages in this still unregulated world of radio. For example, the Houston station was assigned the call letters "HO," Galveston was "GV," and Port Arthur aired as "RA"[36] until 1910 when it was changed to "RU." But, as we shall shortly see, these security measures did little for privacy when equally equipped outside parties navigated the airwaves and quickly learned to intercept and decipher these messages.

Chapter Two

After visiting the Houston station and learning all he could about radio communication from the De Forest technicians, Alfred Daniel went home to start building his own radio sets. Born in 1889 in Austin, Texas, Daniel moved to Houston with his family when he was four.[1] The Daniel family eventually settled just west of downtown in what was then considered the suburbs, in fact, only street blocks away from the bustling downtown. Except for a stint as an enlisted man during World War I, Alfred Daniel would never move from that side of town, residing in the southwest Montrose area at the time of his death.

The Houston of the early 1900s where Alfred Daniel spent his teenage years was quite different from the one we know today. Though dwindling in numbers, large Victorian homes still occupied spaces on Main Street in the heart of the downtown center. Some of the highest man-made structures were the steeples of churches long established in downtown. More affluent Houstonians were moving out from the business districts to budding neighborhoods. Much of the city streets were dirt roads with poor drainage. Still, the new noisy invention called the automobile could be seen more often on the streets and was quickly replacing the horse and carriage. Man's growing confidence in the harnessing of electricity was evident by electric streetcars providing mass transportation in Houston's business districts and electric light fixtures or a telephone in the average, upper middle-class Houston household. The young city of Houston, mapped out with streets and situated along the banks of Buffalo Bayou some 65 to 70 years before, was named after Sam Houston, hero of the Battle of San Jacinto in 1836. The growth of Houston during the late 1800s can be characterized as steady and consistent. Some of its larger resources were the railroad industry and the production of cotton and timber from the nearby vast eastern Texas woodlands. The 1900 census showed that Houston surpassed Dallas, San Antonio, and Galveston as Texas' largest city with a population of 44,633.[2]

Several events would influence the rapid growth of Houston during this time. Ironically, one of these changes did not happen in Houston, but 80 miles east in an area known as Spindletop. Oil was discovered there and the boom began. Hordes of oilmen and investors converged on the area to take their chances. Unable to handle the sudden flood of wildcatters and financiers, the nearby city of Beaumont was forced to relinquish to Houston. The eventual drying up of Spindletop and recent strikes in the nearby Humble and Goose Creek communities also helped to make Houston a haven for the oil industry. Although the city's chief resources were still cotton and lumber production, oil was quickly becoming Houston's gold mine. Recent arriver to Houston and wealthy philanthropist Jesse Holman Jones recognized Houston's potential and soon began the construction of high-rise buildings. Between the years 1906 and 1910 ten new high-rise buildings and hotels were constructed under Jones' direction to decorate the city's downtown skyline.

The hurricane of 1900 accelerated the decline of Galveston's population. Many Galvestonians displaced by the storm moved inland to Houston. With the oil boom, the railroad industry became one of the city's largest employers. For years, Houston had natural inland waterways that were used by small barges carrying cotton and lumber out to the Gulf of Mexico. By the first decade of the twentieth century, a movement was gaining momentum in the city to deepen and widen the water channels to transform Houston into a major southern seaport.

On a social level, many Houstonians in 1906 lived what has been described as a serene and simple lifestyle. Of course, television and radio for the masses had not yet been invented, so other means needed to be explored for entertainment. Instead of living rooms, there were parlors for reading and performing on the piano.

Residents would combat the Houston summer heat by sitting under an oscillating ceiling fan or by relaxing on the front porch and visiting with neighbors outside. Those fortunate to own an automobile could take daylong excursions to beach resorts in La Porte, Galveston, or Clear Lake. But on shell-paved roads, the trip often took much of the day. People also enjoyed boat and yacht trips down the Galveston bay years before Houston's port was officially opened as the Houston Ship Channel in 1914.[3] Picnics were popular where women wore Victorian era dresses, hats and petticoats. In public men were inclined to wear suits, even in the scorching heat. Houston also offered entertainment venues and festivals.

Commencing in 1899, an annual autumn festival called No-Tsu-Oh ("Houston" spelled backwards) would hit the city every November to celebrate and romanticize Houston. Similar to a Mardi Gras-like celebration, the week-long festival started with a parade of exotically decorated vehicles cruising the downtown streets and culminated in the arrival of a king called Nottoc ("cotton" spelled backwards) from a mythical kingdom called Tekram ("market" spelled backwards). The king was usually a prominent city businessman whose identity was not be revealed until his public arrival. While World War I was in progress, Houston's No-Tsu-Oh celebration ceased.[4] This was the Houston Alfred Daniel grew up in.

Alfred Daniel was representative of an emerging group composed almost entirely of young males who took an interest in wireless during the early 1900s and started an informal communication network of radio enthusiasts conversing with each other from long distances apart. Known as amateur operators, these boys revolutionized the science of wireless and led it into a whole new direction. Demographics for this group of wireless enthusiasts were almost exclusively; white, young men between the ages of thirteen and twenty-five, from middle to affluent families, most without formal education in electrical disciples.[5] In the first decade of the twentieth century, amateurs began exploring a free and undisturbed realm with their hobby.

While this sub-culture of young men dabbled in the novice invention called wireless, other laymen in Houston recognized a new discipline people would need schooling in, the field of radio communications. In August 1904, Houstonian Walter S. Ezell decided to offer a course in wireless training in his school named the Houston Telegraph College.[6]

Offering instruction in radio communication was a logical step for Ezell who was dedicated to educating pupils in all aspects of wire telegraphy communication for wire and wireless disciplines essentially used much of the same transmitting and receiving equipment. Concurrent to the time the American De Forest Wireless Telegraph Company first entered Texas to install experimental stations in Dallas and Fort Worth; Walter Ezell acquired De Forest-manufactured wireless equipment for his college. The Houston Telegraph College initially installed an operational, small-scaled radio outfit on its campus, which in all likelihood was Houston's first wireless system. The model wireless system basically consisted of two scaled-down radio towers positioned in opposite ends of the campus building. Students had the opportunity to use the systems and practice transmitting and receiving radio messages to develop an understanding of how wireless worked.

Ezell reported that later in 1904 the college would construct a more authentic system as larger radio towers would be erected in different parts of the city. One would be planted in the area of the Houston Heights community and the other in the then unincorporated town of Harrisburg, an area roughly ten miles southeast of downtown Houston.[7]

The year 1906 was a landmark period for radio communication. Wireless was proving to be a new viable method of communication. On Christmas Eve, wireless inventor Reginald Fessenden conducted a radio experiment in Brant Rock, Massachusetts, which historians credit as being the first documented "broadcast" of music and speech. The year also saw the development of a device that would revolutionize the wireless world. Called the crystal set, this apparatus was able to receive radio waves and was a favorite for the amateur operator. What made it so appealing was that it was inexpensive, easy to build and simple to operate. Touching the end of a live wire nicknamed the "catwhisker" to a point on the flat crystal that was placed between two copper contact points would transform an electrical current into a radio wave. Examples of crystal materials used on these radio sets were silicon, iron pyrites and galena.[8] Before 1906, amateurs indulging in the invisible world of the airwaves were mostly men usually disciplined in a field of mechanics or engineering who had the inclination to make a radio set. But the crystal set opened up the radio scene and made it more available for others to explore the airwaves. As a result, amateurs grew in numbers; younger wireless enthusiasts joined their ranks and not necessarily schooled in the art of engineering. It was no coincidence that Alfred Daniel became an amateur operator the same year the crystal set became available to the mainstream culture.

Those who did not have access to a manufactured crystal set were forced to devise their own receiving and sending crystal sets. Discarded objects around the house became items of necessity for the less fortunate amateur. Old, worn out umbrella stems were used as aerials to receive radio signals. Tin cans and trashed copper wiring wound on broken baseball bats became part of the receiving set. Amateur operators from later generations would remember that cylinder shaped Quaker Oats containers were used instead of baseball bats to wind copper wiring. Ignition coils removed from automobiles and brass spheres from bedposts were connected to a power source to create a spark gap that could be used as a transmitting device. Detecting the occasional and annoyingly faint radio signals required the use of headphones often made from parts taken from discarded private and public telephones.[9]

This first generation of amateurs grew up in a world where distance and space were being conquered around them. The exciting breakthrough called the automobile was changing the way people would commute. Because of the automobile, suburbs were developing around the cities. In the skies, airplanes were becoming more common and people could now travel a span of miles in shorter time. It is no wonder that this environment would breed a new radio audience. Young boys, many of whom could not drive, were conquering their own realms of distance and space locked in bedrooms or backyard sheds and free from immediate danger on their radio sets. Wireless also provided an opportunity for one to traverse the unknown worlds that could not otherwise be explored. Amateurs could converse with another disembodied "voice" miles and miles away with the greatest ease. Or one could eavesdrop on an existing conversation without detection. A ship at sea or a chat with a professional wireless station provided drama for those who could not otherwise know such a situation.

Quite quickly, the radio sets of amateurs began rivaling those of the professionals in terms of sensitivity and efficiency in detecting radio waves in an open atmosphere. As we shall see, the De Forest stations in this region used the same materials in their equipment that amateurs could find around the house and duplicate for their radio sets. Consequently, complaints grew as professionals argued that amateurs were abusing the airwaves with their numbers and their tendencies to disrupt commercial traffic over the air. The fact that no restrictions were made for the airwaves made these professionals nervous. Their concern was that an uninvited amateur could do damage by interfering with a communiqué and possibly send out a false message.

A case in point occurred in 1907 when two amateur operators in Galveston gave "fits" to the De Forest stations in the vicinity and proved that their equipment equaled that used at the professional stations, a situation most certainly duplicated in many other cities. One day in 1907, fifteen-year-old wireless enthusiast George Roy Clough flew a kite with 200 feet of iron wire attached to it with the hopes of picking up radio signals from any station.[10] Clough led the air-borne wire, which was to be used as an antenna, through his window and connected it to his crystal set receiver. After a few minutes of searching for a sender, Clough heard faint radio signals followed by a New Orleans station that came in loud and clear. To the amazement of the young Galvestonian, it did not end there as he heard other local De Forest stations sign on: "Then the Galveston station (encrypted as "GV") called Houston ("HO"), and I heard Houston answer Galveston. Afterwards, I heard several stations working."[11]

Not wishing to stop there, days later Clough employed the assistance of his friend, fellow wireless enthusiast George McKenzie Douglas, to devise a sending radio device. Douglas will later resurface in this history to play a small role in the Houston broadcasting craze of the early 1920s. Using unlikely items such as an old hairpin box, a tin can, an iron pipe, Leyden jars used to store an electrical current and an induction coil to transmit the spark that sent the radio signals, the boys developed a set that duplicated those used by the De Forest professionals. On one occasion George Clough was able to interrupt transmissions between two De Forest stations and send out his own message, causing an infurious reaction from the local professional wireless operators.

Following this incident, Clough received a telephone call from the operator of the Galveston station warning him not to interfere with transmissions again. By this time it seems that Clough had already gained quite a reputation with the local wireless operators. On any given day, Clough was able to send messages and copy the stations in Houston and Port Arthur. The furthest he was able to receive was 150 miles out in the Gulf of Mexico when the Mallory Line passenger steamship, *Concho*, sent a message to Galveston on a Sunday afternoon in 1907.[12] Like Alfred Daniel, who was an amateur in Houston during the same time, Clough would make radio a career later in life, working at a Beaumont broadcasting station in the 1920s as a technician, followed by operating the broadcasting station KLUF in Galveston before finally becoming mayor of that city in 1955.[13]

Tensions between Lee de Forest and Abraham White reached a breaking point in the summer of 1906, the time the company began launching a series of stations in Texas. Following the court's decision to have the company pay damages for the Reginald Fessenden infringement suit, it was also revealed that the company had attained a huge debt against its creditors. The wireless company named after de Forest was only in name now for the wireless inventor had little to do with the everyday activities of the company. He never personally inspected the minor success his system had in connecting Houston, Port Arthur, and Galveston through wireless.

De Forest had enough problems of his own to handle. White began to staunchly ostracize him from business activities and publicly blamed him for the Fessenden lawsuit problems. By the time the Houston station was erected, de Forest had reached a fateful decision. In November 1906, de Forest announced his resignation from the company and White granted him a severance pay of $500 and, more importantly, released the rights to an invention de Forest had made during his employment with the company. Perceived as useless by White, this invention would save the career of de Forest and prove to be one of the most important building blocks of radio technology and influential in many fields of electronics.

Following de Forest's resignation and the lawsuits, the American De Forest Wireless Telegraph Company had to change some of its practices. However, one thing that did not change was the company's ability to attract controversial characters. One of these was H. H. C. Dunwoody. Prior to his involvement in wireless, Dunwoody was stationed in Cuba in 1900 as part of the United States Weather Bureau. Erik Larson's book on the Galveston hurricane of 1900, *Isaac's Storm*, portrays Dunwoody as an immoral and shady individual. Conspiring with others in the Bureau for vain reasons, Dunwoody prevented the sending of wire telegrams to Texas to warn of an approaching hurricane that struck Galveston, killing thousands and known historically as the nation's worst natural disaster of the century.[14]

Dunwoody's role in the American De Forest Wireless Telegraph Company became crucial when Abraham White needed a new receiver for his wireless system due to de Forest's resignation and Fessenden's lawsuit of 1906 that prevented White from using the one he had. Dunwoody, current vice-president of the company and one time member of the United States Signal Corps, developed an inexpensive, sensitive and rugged wireless receiver that the company adopted. All stations in Texas were equipped with this receiver made from crystal carborundum, a compound of carbon and silicon and used prominently in amateurs' crystal sets.[15]

Lawsuits and poor financial planning placed the American De Forest Wireless Telegraph Company on the verge of bankruptcy. To alleviate this, White devised a plan that was legal under Bankruptcy laws. Under a legally valid stipulation, the company's debts and liabilities were dissolved when White reorganized the company to become the United Wireless Telegraph Company. Then, White also saw his authority over the company quickly vanish.

In the summer of 1907, vice-president Christopher "Columbus" Wilson succeeded White's presidency after some scorching bad press was printed on him.[16] Eventually, the Houston station became a victim of its marginal usefulness and of the company's debt obligations. Without any newspaper announcements, it appears that the station was dismantled sometime in 1908. The other stations the company claimed it had erected in Dallas, Austin and Waco never appeared in any government listing, which indicates that they never functioned. By 1909, all De Forest stations in Texas were removed except for those in Galveston and Port Arthur.[17]

Finally, the federal government intervened after complaints were made on these unethical wireless ventures. Its efforts reached a climax when in 1910 it began serious investigations on these wireless companies and brought charges against and arrested the high officers of the De Forest company, including White and Wilson. As part of the judgment, the Galveston and Port Arthur stations that continued to function were assumed by a Marconi wireless company to serve maritime traffic in the Gulf of Mexico.[18] White and Wilson were convicted of mail fraud for their solicitation of fraudulent stock through the mail service. Both served prison sentences and eventually faded into wireless history. But this would not be the end of corrupt companies entering Texas to sell the wild aspirations of wireless to gullible investors.

Spared the judgment of the convictions was Lee de Forest, who was able to escape the decision and continue his interest in wireless. De Forest's hopes and dreams relied on one the few things he was allowed to take with him following his parting with Abraham White. His idea was to market an improvement on an invention he devised that would enhance clear transmissions of voice and music through radio signals, the concept known as radiotelephony, and through a slightly modified radio set called the radiotelephone or wireless telephone. Called the audion, de Forest's invention resembled an incandescent light bulb and was an improvement to inventor John Ambrose Fleming's glass bulb detector of radio signals. De Forest bettered Fleming's idea by adding an extra element to his device that considerably improved and amplified the reception of radio signals. Although not as reliable and powerful as wireless telegraphy, radiotelephony would be the next advancement in radio communication.[19]

Originally, de Forest conceived of marketing his audion in connection with point-to-point communication. But shortly after his resignation from the American De Forest Wireless Telegraph Company, he saw the potential of reaching out to the ever-growing number of amateur radio operators searching the airwaves for amusement purposes. Starting in 1907, de Forest began demonstrating his radiotelephony equipment to interested parties such as the United States Navy.[20] Along the way, he successfully conducted wireless tests of transmitting speech and music to a scattered "audience" of amateur operators. One of these historic "broadcasts" came from de Forest's New York office and aired music that was heard by several local, amateur operators. It is here that historians have credited the radio inventor with conceiving the idea of radio broadcasting as we define it today, even though months before, his rival Reginald Fessenden, aired the first verifiable transmission of music and speech on December 24, 1906, from Massachusetts. Equipped with radio sets from Fessenden, ships out in the Atlantic Ocean reported hearing the Christmas Eve "broadcast" containing Nativity music and a short lecture from Fessenden himself.

While radio broadcasting was still fresh in his mind, de Forest traveled the nation hoping to successfully market his radiotelephone for more practical reasons and to offer a legitimate rival to wireless telegraphy. Similar to the principle of modern cellular phone technology, his concept would transmit voice communications between two parties. Unfortunately once again, he fell in with the wrong people to help market his scheme. Joining him was James Dunlop Smith, a former stock salesman from Abraham White's American De Forest Wireless Telegraph Company who helped the radio scientist organize the De Forest Radio Telephone Company in 1908. The wireless inventor was very enthusiastic as he attempted to peddle his radiotelephone to all parts of the nation. From the corporation of the De Forest Radio Telephone Company came a subsidiary that considered Houston as a possible site for one of its radio stations. Called the Atlantic Radio Telephone Company, this company came to Houston in the spring of 1909 to announce the opening of a station that could transmit voice messages.[21]

As the young lumber magnate named Jesse Jones started taking a fancy to constructing buildings for Houston's skyline, the Atlantic Radio Telephone Company contracted with him to install one of their stations on top of the *Houston Chronicle* building, which stood nine stories high. Earlier, the wireless company had secured agreements with other coastal cities, including Galveston, to establish a chain of communication links. Some of the intended list of potential investors included those wealthy enough to navigate the bays and waterways, which was reflected in an article written about the proposed station in Houston by the *Houston Chronicle*. Houston mayor Horace Rice had recently purchased a yacht called the *Zeeland* and used it as a promotional vehicle to encourage Houston to acquire access to the Gulf of Mexico by constructing a deep channel inland port. "The mayor could make decisions from his yacht out in the gulf as if he were in his office,"[22] was one of the bold predictions made in the *Houston Chronicle* newspaper as it announced the placement of de Forest's radiotelephone station on its new building.

But what could have been the city's first radiotelephone station turned out to be a false dream. The station was never even constructed. Even had it reached completion, the efficiency of the station would, in all probability, have been poor. Like de Forest's former company, the plan was saturated with fraud. Showing little concern for the advancement of wireless technology, the company's officers instead pocketed the profits from inflated stock sales and gave little for development of the de Forest radiotelephone. When all was said and done, most of the company's stations failed to work altogether and the few that operated did so inefficiently.[23] Judgments and convictions were given to some of the company's top officers.

Incredibly, once again, de Forest was spared judgment, giving him yet another opportunity to launch a new venture based on his radiotelephone. Because so much controversy had surrounded de Forest in the Northeast, the radio scientist moved to the northern

Pacific coast where he based his next, and final, radio company. Called the North American Wireless Corporation, the company announced in 1910 that de Forest would personally demonstrate and erect stations in Houston and other Texas cities.[24] However, by this time, competition was fierce in wireless technology as another company invaded Houston to install a radio station. Thus, poor business decisions from de Forest prevented him from getting his new company off the ground.

Lee de Forest would continue to conduct more historic demonstrations on his radiotelephone before World War I and would eventually leave radio experimentation to, among other things, concentrate on producing audio for the then silent motion pictures industry.

By the early 1910s wireless technology still had not developed enough to outdo the overland communication systems of the telegraph and the telephone. A more sensible use of wireless was as a backup to existing communication lines and a few smart individuals realized this. Owner of Houston's prestigious Rice Hotel, James Lawlor, installed a wireless set in that hotel in January of 1908. Containing all the modern amenities of its time, the Rice Hotel, nevertheless, had to deal with an occasional inefficient telephone system. The wireless sending system was described as "a tiny square box with two upright movable arms to which are attached two brass knobs about the size of golf balls." When the system was ready to send, it "sang like a giant June bug" and threw out a "five inch long flash of lightning."[25] An aerial serving as the antenna for the radio apparatus was mounted on top of the hotel building. The system was said to have a range of 50 miles and could cover the distance to a 10-foot antenna mounted on the windmill of James Lawlor's farm out on the pasture fields on Dairy Ashford Street.[26]

The historic oil strike in Spindletop in 1901 brought many men to southeast Texas, including one wildcatter named Joseph S. Cullinan. A year later, he and a few of his wealthy friends formed the Texas Company in Beaumont, Texas. Later to be known as Texaco Incorporated, the Texas Company was an oil-producing and oil-transporting corporation that, from the start, maintained an aggressive policy of transforming itself into a self-reliant company. In short time, the company acquired lands rich in crude oil and built extensive overland pipeline systems that connected its facilities to the nearby seaport of Port Arthur, Texas and the inland town of Shreveport, Louisiana.[27]

Recent oil strikes in the nearby Humble field and Goose Creek area, with assurances from Jesse Jones to construct a building for the company in Houston, convinced the Texas Company to move its general offices from Beaumont to Houston in 1908. Waiting for the company was a brand new 10-story building built by Houston's favorite philantropist.[28] Continuing its policy of being a self-sustaining body, the Texas Company already had a small, independent fleet of oil barges transporting the company's most important product when it moved its headquarters to Houston. High tariffs imposed by railroads and pipelines from other entities to transport oil made the Texas Company realize that its maritime affairs were a crucial component to its enterprise.

The need for instant communication was also perceived as a vital element for the Texas Company. A common problem plaguing oil companies at this time was the threat of a large oil fire to a company's pipelines. Access to immediate communication with an area strickened with an out-of-control fire could be the difference between losing million of dollars in crude oil or coordinating an effort to save an important pipeline system. However, during a state of emergency, public telephone lines and even telegraph lines were often congested with concerned parties trying to contact the area or, even worse, downed by the inferno.

For this reason, the Texas Company constructed and maintained its own private telephone and wire telegraph lines to have instant contact with its outlying areas in times of emergency. With improvements in radio technology by 1909, the company contemplated the possibilities of installing wireless stations to serve as a backup means to communicate. In some cases, wireless was the only means of communication in desolate and remote locations where the company had pipelines with no wire communications in place. Coordinating a growing fleet of oil barges also became a dilemma for the Texas Company. Often, oil fires ablaze on a barge out at sea required a quick rescue response effort that could only be facilitated by wireless communication.

What seems like an unlikely pioneer in radio technology, the Texas Company began experimenting with wireless. During the time it moved its headquarters to Houston, the company installed wireless outfits in Shreveport, Louisiana, and in Port Arthur, Texas to remain in contact with its ships. Then, as the company's marine section grew and became a separate department in October of 1910, a huge radio tower was unveiled atop the Texas Company's Houston building.[29]

Measuring sixty-six feet above the ten-story building, the tower had extending from the tip of its mast an aerial wire that stretched to the top of the garden roof of the Bristol Hotel located across the street. The Houston station completed the tri-circuit system to complement those already in operation in Port Arthur and Shreveport. The three stations were kept in readiness on a daily basis and proved to be the area's most efficient overland wireless system at the time. In another year, wireless equipment would be outfitted on most of the company's marine vessels. By 1912, the company's wireless endeavors would expand to include inland stations into Oklahoma.[30]

When in operation, Houstonians passing by the Texas Company building, many that had never seen wireless in operation, could hear electricity crackle from the wireless station and see blue flame sizzle from the mast on top of the building. Eventually, the Houston station as well as Port Arthur's station directed the traffic of messages sent and received from the ships out in the Gulf of Mexico.[31]

Wireless grew in popularity in Houston by 1910, as it never had before. More young boys were becoming amateur operators, the Texas Company completed its wireless station in Houston and another company would enter into town to prey on the public's naiveté on wireless. That year also saw what might have been the beginnings of commercial advertisement through broadcasting in the city. As mentioned before, de Forest thought of broadcasting in early 1907. His isolated radio broadcasts continued into 1910 including an airing of a live opera from New York and political speeches from his mother-in-law. While these broadcasts may not have reached Houston amateur operators, evidence shows that the notion of reaching many people through wireless may have been on some Houston minds in 1910.

On July 10, 1910, a half-page advertisement was published in the *Houston Chronicle* and the *Houston Post* for the two hugely popular department stores of the Mistrot-Munn Company. Heading the ad was the phrase, "A Wireless Message from Mistrot-Munn Company, 'Buy Now and Save in Cost.'[32] The ad shows an illustration of a young man on headphones dictating a message he is receiving through radio waves. Given the circumstances, it seems very plausible that owners of the department stores would equip their establishments with a wireless system and attempt to capitalize on the latest craze. The Mistrot family had a chain of stores throughout the state and was known for its imaginative marketing schemes. In the months of April and May in 1897, newspapers cited that many Texans reported sightings of strange, unknown flying machines and the phenomenon has gone down in Texas legend as the Great Airship Mystery. During this period, the Mistrot stores in the Dallas area ran several clever, illustrated newspaper ads integrating the airship mystery incidents with its advertising ploys.[33] It was similar to what the Mistrot-Munn stores were achieving with wireless in 1910.

Though the Great Airship Mystery proved to be composed of false, exaggerated stories perpetrated by newspapers to seek more readers, the notion of Houston merchants utilizing wireless stations in 1910 for advertising gimmicks was not far fetched. Around the nation, other department stores maintained radio stations and later held some of first broadcasting licenses of the early 1920s. Legends concerning a tragic, historic event grew from a store that had installed a station in 1911 or 1912. The Wanamaker stores of Philadelphia and New York had stations at both stores for intercommunication and publicity purposes.

In July of 1910, this advertisement appeared in the local newspapers for a well known department store. Quite obviously, wireless communication was already in existence in Houston between amateur operators. *Courtesy of Houston Post archives, Houston Public Library*

It so happens that the operator was a young amateur operator named David Sarnoff who, on April 12, 1912, picked up faint distress signals from a sinking passenger ship called the *Titanic* on her maiden voyage. For hours on end following his initial contact with the

Titanic, Sarnoff remained at his equipment receiving up to date information of the tragic event for concerned relatives and friends crowding the New York department store. The event made Sarnoff an instant household name and led him to a historic career in radio.[34]

With a Mistrot-Munn store at the corner of Congress and Main Street and another street blocks away on Travis Street, it seems logical to deduce that these imposing department stores maintained wireless stations by 1910. Some twelve years before the term "broadcast" was coined, inventive Houston merchants may have been advertising over the airwaves. But evidence would show that this was not an isolated example.

A month later, another peculiar ad appeared in the local newspapers. An illustration of an aerial with radio signals radiating from its tip was shown with the phrase, "A Wireless toYou."[35] The ad was for J. W. Thiel, a local contractor specializing in plumbing and lighting jobs. Since Thiel was an electrician, he was likely familiar with wireless and may have even been an amateur operator. Thiel may have also been an essential supplier to local wireless enthusiasts building and maintaining their own radio apparatus.

It should be noted that these ads implying radio transmissions provided no prearranged frequencies or any schedules as to when they would air. No other evidence exists of these early ingenious advertising schemes, only their mention in newspaper ads. In examining these advertisements one can see clues that wireless was heading into another direction by 1910. Little did anyone realize that these imaginative merchants were unintentionally providing a glimpse of a new dimension in radio communication called broadcasting.

A month after the Mistrot-Munn Company "wireless" advertisement, local electrician/plumber J. W. Thiel placed this ad for his establishment. Were certain, local entrepreneurs "broadcasting" their services and goods over the radio waves some twelve years before the term "broadcast" was coined? *Courtesy of Houston Post archives, Houston Public Library*

Chapter Three

John B. Dickinson came to Houston in early 1911 to win support for a new wireless company, the last effort to transform his venture into a reputable organization. Addressing the Houston Chamber of Commerce on January 30, 1911, Dickinson hoped to gain interest in constructing a station equipped with a powerful 25 to 35 kilowatt transformer in the city.[1] Such a station could provide instant communication to New York and ships far out to sea, an enticing offer made to a city beaming with enthusiasm as it continued to widen its ship channel and turn Houston into a major seaport. A week prior to his engagement with the Chamber of Commerce, Dickinson gave a lecture on the principles of wireless at the Houston YMCA followed by a talk at Texas A&M University where he installed a station on campus so students could experiment with.[2] The company Dickinson supported eventually became known as the Texas Wireless Telegraph & Telephone Company, a fiasco of an enterprise that marked the decline of the fraudulent wireless companies in Houston.

Formed around January 1, 1910 in San Antonio,[3] a financial institution called the Texas Fiscal Agency planned to finance a wireless commercial venture. To test how Texans would respond to a wireless company, the Texas Fiscal Agency initially began selling stock of such questionable companies as the Collins Wireless Telephone of Newark, New Jersey and the Continental Wireless Telegraph and Telephone Company of New York. Shortly after this, it was decided to create a subsidiary of these companies to serve Texas. Before a charter was drawn up for the corporation, the agency began introducing stock for the wireless company with claims that it would be owned and managed by Texans exclusively.[4]

Patterned similarly to the De Forest companies of past years, this company proposed a chain of stations throughout the Lone Star State to provide wireless service. Recent improvements in wireless telephony offered an added enticement to lure and exploit new prospective investors. Still not as reliable in long distance communications as wireless telegraphy, the company proposed to install smaller stations to serve as relays along the chain of stations to provide the radiotelephone service. For this service, the company decided to use the equipment of A. Frederick Collins, an early experimenter who had achieved some success with sending voice transmissions over radio waves. This company that was to have a monopoly of wireless service in the state was initially called the Texas Collins Wireless Telegraph & Telephone Company.[5] Heading the brokering of this company was John Dickinson, a man claiming to have been connected with wireless for about ten years. Another broker of the Texas Fiscal Agency was William Bomar who was recruited by John Dickinson and was a real estate broker operating near Houston before his employment with the agency. Before this, Bomar had no prior experience in wireless.[6] The proposed wireless company was presided over by G.R. Spielhagen[7] and was headquartered in San Antonio. The Texas Fiscal Agency underwrote the business by issuing stock to the public and gave itself the exclusive right to sell the wireless stock. As specified by the agreement, sales of the public stock would establish a charter and purchase wireless equipment. In time, more income would be generated as a commercial service was established.

An initial capitalization of $2,000,000, divided in shares at $1 each, was established. The Texas Fiscal Agency was given unlimited means to sell the company's stock with an agreement that the agency would earn any gain it achieved above the par value.[8] It turns out the agency made a hefty profit from this arrangement, often selling the wireless stock three times above the $1 per share. By April 1910, the agency began introducing the stock to the market with an aggressive stock selling campaign. Among its marketing schemes to sell stocks in this new business venture was the distribution of brochures describing the bright future of wireless communication through the postal service.

Advertisements began appearing in the newspapers for the sale of stock. A local representative of the Texas Fiscal Agency began running ads in April in the Houston newspapers that read:

Collins Wireless Telephone

Parent stock. Best investment on earth. I bought when it was cheap and will dispose of a limited amount at $3 per share and will have stock transferred directly to you.[9]

Before a single station was even erected and no formal incorporation of the company was drawn up, inexperienced stockbrokers were going to the public with an unrealistic proposal using equipment manufactured by questionable companies. The recent history of deceitful wireless companies should have warned potential investors that this Texas enterprise was something to avoid. Still, some Texans saw the wireless company as a solid investment.

One of these was Texas resident Zacariah T. Goolsby who was approached by a representative of the agency and purchased 100 shares in the proposed company. Agreeing to invest $250 for his shares, Goolsby later realized that the deal was questionable and decided to back out. Wishing to have a refund on his investment, the Texas Fiscal Agency denied the return of his money pointing out a clause in the contract that initial stock in the venture was "non-negotiable and non-transferable two years from the date of purchase."[10]

No outside, regulatory party was in place to monitor proceeds gained from investors. Records may not exist to accurately reflect the profits made by the Texas Fiscal Agency. But one can estimate that during this period when soliciting wireless technology was still unregulated, the brokering house made a substantiate amount of money.

In July 1910, Texas Fiscal Agency manager John Dickinson wrote an article in a monthly publication called the *Southwestern Electrician* detailing the company's plans to install a wireless system throughout Texas. The article gave him added public credibility that his venture was legitimate and even cited the areas in Texas where his wireless system would service. Cities like Houston, San Antonio and Galveston would be equipped with stations powerful enough that communication between them through radiotelegraphy would be done with ease. With an ever-improving radiotelephony field developing, Dickinson also enticed potential investors with an assurance that Texas would have a radio service able to transmit speech messages. Admittedly, since radiotelephony was a more difficult form of radio communication to transmit with, Dickinson contended in his article that the radiotelephone service would only be effective by placing smaller stations along the installed radiotelegraphy lines to serve as relays. Victoria, Fredricksburg and Wichita Falls were cited as possible locations for these radiotelephony relays. As others did before him, Dickinson pointed out that when an effective wireless system was in place it would competitively rival wired land communications in quality and service rates.[11]

By the end of the summer, funds continued pouring in as agents scoured the state searching for more investors to commit to this wireless scheme. But not so much as one station was in operation with no convincing indication that the system would work. Radiotelegraphy had already been an established science with a proven record of sending messages hundreds of miles. The real test, for the company, would be to convince investors to buy into the wireless telephone system. Before 1910, wireless telephone demonstrations had been scattered. The events from Reginald Fessenden's historic music broadcast in 1906 to Lee de Forest's documented experiments in transmitting speech and music proved that radiotelephony was a reality. Other companies also claimed success with the wireless telephone but none had established a commercially viable long distance system before 1910.

For demonstrations of their wireless system, the Texas Fiscal Agency relied on equipment to be sent from one of its parent companys in New Jersey. However, it seems that Dickinson and the Texas Fiscal Agency were also victims of the fraudulent wireless schemes running rapid. Promised equipment was not delivered.[12] Already on a course where breaking with investors was unacceptable, the Texas Fiscal Agency needed to quickly acquire wireless telegraphy and telephony equipment. Radiotelegraphy equipment was somewhat easy to make or acquire. The more difficult task would be to locate a functional radiotelephony system. Knowing full well that a system transmitting voice messages over long distances was still not feasible, the agency turned its attention to locating an inventor that had made a small-scaled radiotelephony apparatus to satisfy the investors. That inventor was found in a seventeen-year old amateur operator.

Bernays Johnson, son of former Missouri governor Charles Johnson, was invited to Texas to manufacture his design of the wireless telephone. Said to have experimented with wireless transmissions of voice since the age of seven, Bernays Johnson finally had invented a working model in his teen years.[13] Anxious to present an operational wireless telephone, the company purchased the rights to Johnson's invention. Commencing in the summer of 1910, the company made public their radiotelephone. A "toy" model was installed in the *Houston Chronicle* building where the Houston office of the company was situated. By September, editors of the *Houston Chronicle* were invited to test the system. Transmitting voice messages over a distance of several floors in the building, the system was meant only to demonstrate and sell more stock. A radiotelephone on the second floor was tuned with a wireless receiver on the ninth floor. Reception quality was judged as excellent as even the ticking of a watch could be heard over the airwaves.[14]

Another example of demonstrations staged for the company's wireless telephone occurred in Victoria, Texas in August 1910. The coastal town was chosen as a site for Houston's relay station and a campaign was started to finance it. Known as "Acquaintance Day," August 5 was set aside as a day for Victoria citizens to get to know new companies entering the area. One of these new businesses called itself the Texas Collins Wireless Telegraph & Telephone Company, though the equipment originated from Bernays Johnson and not A. Frederick Collins. Officers from the company came from San Antonio to stage the radiotelephone exhibit. At the local courthouse and "witnessed by large crowds,"[15] the radio apparatus was demonstrated in periodic sessions throughout the day. Recorded accounts of the demonstrations, though, did not say what distance the radiotelephones covered or if music was also transmitted. But during this time when radio was still in its infancy, the average Texan witnessing any kind of transmission of voice over radiowaves must have been impressed enough to think that wired communications would soon become an obsolete form of long distance conversing.

Riddled with controversy and failures, investments still continued arriving into late 1910. Dr. J. G. Baldwin was persuaded to buy into the company in October. Thinking he was purchasing 1,000 shares of stock, he gave the agency $1,000. Much to his surprise, he was handed a contract for only 250 shares, illustrating the fact that for every $1 going to the establishment of the wireless company, $3 were awarded to the Texas Fiscal Agency. In a later meeting, Dr. Baldwin approached Dickinson, hoping to get a refund. Like other investors though, he was denied due to his ignorance of the clause printed on the back of the shares stating that the investment was initially nontransferable and nonrefundable.[16]

With the recent convictions of Abraham "Lincoln" White and Christopher "Columbus" Wilson in connection with Lee de Forest and the United Wireless Telegraph scandal, the federal government continued its offensive against questionable companies involved with wireless. On November 21, 1910, U. S. postal inspectors raided the offices of the Continental Wireless Telegraph and Telephone Company in New Jersey. This was the organization that had introduced its stock into Texas through the Texas Fiscal Agency and later agreed to sponsor the company now organizing as the Texas Collins Wireless Telegraph & Telephone Company.

Heading this raid was William E. Greenway, a postal inspector in charge of these investigations. Under his jurisdiction, Greenway alleged that wireless companies around the nation were soliciting falsely leading material in the mail to sell stock. It was probably during this investigation of the New Jersey company that attention was drawn to the Texas Collins Wireless Telegraph & Telephone Company. Continuing on his crusade, Greenway quietly arrived in Texas in early 1911 to examine the dealings of the Texas wireless enterprise and readied himself to conduct a raid on the Texas Fiscal Agency.[17]

Finally, after nearly a year in operation and months of attracting investors, the Texas Fiscal Agency began constructing radio stations and chose San Antonio to place its first station. With the pilot station installed in San Antonio, the nearby town of Fredericksburg was chosen to place a relay station for it. By the end of 1910, the two stations were said to be in working order and in "commercial" availability for transmitting telegraphic messages. At the exorbitant rate of twenty-five cents a word, messages were transmitted during the daytime. Because successful transmissions were more often achieved at nighttime, the going rate to send messages was markedly reduced to fifty words for twenty-five cents for evening transmissions.[18]

Meanwhile, permits were secured to install the Houston station on the newly constructed Carter building in downtown. Named after its owner, lumber and banking businessman Samuel Fain Carter, the sixteen-story building created a stir in the city when it was constructed. Prior to 1910, Houston's downtown skyline was almost single handedly created by Jesse Jones. Impressed by the architectural landscapes in some European cities, Jones envisioned a Houston skyline with buildings not having more than ten stories and he had been able to maintain a hold on the Houston building scene until Samuel Carter decided to make a building. Against Jones' preferences, Carter constructed the sixteen-story edifice at the corner of Main and Walker Streets.[19]

The Carter building was completed in early 1910, exceeding any other Houston structure by at least six stories. It was on top of the Carter building that the Texas Collins Wireless Telegraph & Telephone Company agreed to place its station. The station resembled a penthouse with the antenna tower rising above it and wires stretching from its mast across the street to the roof garden and eatery of the Bristol Hotel.[20] The Carter building was located across the street from the Texas Company building that had also recently mounted a radio tower above its location.

Completed and functional, the Houston station was said to be operational in January 1911. Press releases from the Texas Fiscal Agency later retracted this statement and reported that the station would not be ready for commercial use until its companion station was operational in Victoria, Texas.[21] That would not occur for another four months. A more likely scenario is that in its rush to implement a statewide wireless system, the Texas Fiscal Agency failed to acquire enough personnel to operate the stations, a problem that perpetually plagued the company. February saw the opening of a wireless outfit in Waco and, immediately, Houston established communication with it. Shamelessly, the communication between the two cities was established by chief electricians who were not employed with the wireless company but with the municipal bodies of Waco and Houston. In charge of all electrical endeavors involving the City of Houston, Clarence R. George acted as the local station's wireless operator when Houston reached Waco on February 20, 1911. After wireless courtesies were exchanged that day, George ended his message with the comment, "Wireless sounds good to me."[22]

Radio tower atop the Texas Company's building in 1910. In the background can be seen construction in progress for a wireless tower for the Texas Wireless Telegraph & Telephone Company atop the Carter building. *Courtesy of Special Collections & Archives, University of Houston Libraries.*

Successfully putting the Houston station on the air gave Dickinson the necessary boost to launch a new wireless campaign. The recent arrests of its parent company in New Jersey probably convinced directors of the Texas Collins Wireless Telegraph & Telephone Company to downplay its failed wireless telephone system and concentrate on long distance radiotelegraphy instead. Dickinson now pushed the implementation of a much larger station in Houston, one strong enough to have direct communication with those as far away as New York.[23] His exhaustive efforts led him to address Houston's Chamber of Commerce, the local YMCA society and the Texas A&M University in College Station.

But the bubble would burst for the Texas Fiscal Agency on March 1, 1911, when Greenway and other postal inspectors raided the offices of the agency in San Antonio. Dickinson and William Bomar were arrested on charges of mail fraud. Although the two posted bond and attempted to downplay their arrests, the damage proved to be irreversible. Coincidentally or not, about this time, the wireless company had finally received its charter to conduct business in Texas and was now known as the Texas Wireless Telegraph & Telephone Company. Since Dickinson and Bomar were associated with the Texas Fiscal Agency, the new wireless company attempted as to save face as much as possible, severing its ties to the doomed agency.[24]

Most likely delayed by the arrests of Dickinson and Bomar, a station was eventually opened near Victoria, Texas on May 6, 1911. That day, messages were exchanged as far away as San Antonio. As a friendly token to settle any controversy, the Texas Wireless Telegraph & Telephone Company's president, G. R. Spielhagen, offered Victoria residents and enterprises ten days of free service to transmit messages. He also received surprising endorsements from a local Victoria newspaper. On the day the station opened for business, the *Victoria Daily Advocate* wrote, "So, then let us be sensible and conservative and slow to judge in all things and not ready to decry something that we cannot understand."[25] During the promotion, some entrepreneurs did indulge in transmitting messages. A merchant in

Victoria authorized a bill of goods for fruit to be sent over the airwaves to the Houston station for a local produce farmer.[26]

Unfortunately for the company, the damage had been done. The stock investments ceased to come in, thus leaving the wireless company to support itself.

The limited amount of stock investments that eventually reached the wireless company did produce a service, though far less efficient than what was originally promised. Under extreme, favorable atmospheric conditions, the Houston station reportedly was able to send a message about 5,000 miles away. The station on the receiving end also proved to be effective. It was able to receive messages from as far away as Norfolk, Virginia and a station situated on the southeast coast of Cuba.[27]

In the end, only a handful of the company's proposed stations reportedly performed with any efficiency. The two larger units in San Antonio and Houston of course signed on, and the smaller, relay stations in Fredericksburg, Waco and Victoria also reported activity.

From 1911 to 1913, a black cloud covered the wireless company while the Texas Fiscal Agency was on trial. What was revealed was no surprise to those having experience in these wireless affairs. Thousands of dollars had been swindled from investors around the state. The radiotelegraph was functional, but the promises of installing a radiotelephone service fell short. In fact, no large-scale telephone service was ever operational. Demonstrations, such as in Victoria during "Acquaintance Day" were small, covering a distance of maybe only a few street blocks at the most.

After a year of trials, John Dickinson was found guilty of mail fraud and received a sentence of two years. William Bomar, the agent in charge of promoting the radiotelephone system, was also found guilty and given a sentence of eighteen months.[28] Quietly going into the obscure pages of Houston history was the Texas Wireless Telegraph & Telephone Company. Sometime between

1913 and 1914, the huge tower above the Carter building went down. The dark, embarrassing years of wireless had come to an end.

Following this period, the local print media must have realized that it had also fallen victim to these unrealistic claims made by the fraudulent wireless enterprises. Instead of inquiring for more in-depth information regarding a budding wireless technology, newspapers often accepted statements from these scheming individuals at face value. Disheartened, newspapers made a conscious effort not to feature wireless progress for the duration of the 1910s. Wireless-related subjects were mostly reported in trivial articles and placed in the back pages of newspapers. Years would go by before the print media would again devote a substantial amount of its attention to radio.

Chapter Four

Following the con artists and the scams, the emerging dominant wireless group in Houston was the amateur operator. The young boy tinkering with radio signals locked away in his room or in a backyard shack became the leading catalyst unknowingly directing the destiny of radio. Amateurs such as Alfred Daniel who already had acquired experience were recruiting new, younger members into this informal fraternity. Born at the turn of the century, these new amateurs had the availability of better equipment. Crystal sets, the poor man's radio receiver introduced in 1906, were now being perfected and mass distributed by radio manufacturers and dealers anticipating the rising popularity of amateur wireless activity in the 1910s. Pioneers like de Forest, who had survived the bankruptcy and convictions of his wireless companies discovered a new market for his audion in those amateurs fortunate enough to acquire the voice-detecting and transmitting, clear, bulb-like device.

Unbeknownst to them, these young men and boys spending hours in the disembodied world of the ether were trailblazing new ground and setting the foundation for radio broadcasting. In their efforts to find new ways of entertaining themselves, these amateurs covered much of the formats composing radio broadcasting today, especially after World War I, when they began experimenting with radiotelephony to scatter their transmissions to an audience of many listeners.

But the road for these amateurs would not be a smooth one. Obstructions would be thrown in their way, almost exclusively from the federal government. The relationship between the amateurs and the government had always been on shaky ground and in the years of 1910-1912 as major crackdowns on wireless ventures were being conducted, the government attempted to regulate the airwaves. The Radio Act of 1910 was passed that required all ships carrying a certain number of personnel to be equipped with wireless equipment and obtain a license to operate the said radio set.[1] Pressure from private, industrial groups as well as government operators created the Radio Act of 1912 that directly affected the amateur. According to this legislation, all amateurs were required to obtain a license for themselves or their radio set, regardless if they were transmitting messages or simply listening to other conversations over the air. This Act also restricted amateurs to operate at the radio frequency of 200 meters (1500 kHz) or above. However, the Act also provided that selected amateurs could be granted a special license giving privilege to operate on other frequencies usually allotted to government and private industry wireless operators only.

In times of emergency, amateurs were ordered off the air and forbidden to interfere with any wireless traffic. Furthermore, as specified by the Act of 1912, amateurs could only outfit their apparatus with a transformer no stronger than 1 kilowatt.[2] Government and commercial stations were allowed more powerful transformers. But for all the regulations imposed on the amateurs, it remained relatively easy for a novice to obtain an operator's license. No exams or prerequisites were required. One only needed to ask for a permit from the U. S. Department of Commerce, who would then assign the operator a user license. The Act of 1912 was created to curtail the growing waves of amateurs who were still perceived as nuisances eager to use the ether as an entertainment vehicle. It, however, did not curve the amateurs' use of the air, and its popularity continued to grow.

Even harsh warnings from government authorities did not deter amateurs and their activities. During Europe's involvement in World War I, U. S. customs officials advised that any amateur violating the neutrality of American ships by interfering with their wireless traffic would face serious charges.[3] Persons not holding at least an operator's license to transmit would especially be subjected to criminal prosecution.

In 1913, the government began publishing an annual listing of licensed radio equipment and assigned call letters to the stations. Based on these records one would conclude that Houstonians into radio were few as the first license for a radio station did not appear in the city until 1915, and thereafter, only one other amateur license for a radio station would be granted in Houston before 1917. But nothing could be further from the truth. Amateur wireless activity was alive and well in pre-World War I Houston. Although government documents existed listing the licensed amateur stations, no listing was published accounting the amateurs who were granted operator licenses. Thus, it appears that most Houston amateurs opted to license themselves only and not their stations.[4]

Civil engineer by trade and resident of Houston's Heights community, Charles Edwards was operating at the maximum limit allowed of a one-kilowatt transformer and was sending messages to others 1,000 miles away by the end of 1916.[5] However, no evidence can be found that Edwards licensed his station nor is there an existing publication showing that he acquired an operator's license.

How many amateurs were active in Houston before 1917? That question will probably never be accurately answered. But based on how quickly wireless know-how assimilated through the airwaves as well as the informal network of those interested parties comparing the principles of radio, a conservative estimate would place the number of Houston amateurs at the time in the range of 60 to 80.

To the average citizen, wireless, for all its worth, was still perceived as an eccentric activity practiced by young boys, electrical engineers, and a few college professors interested in its principles. Local newspapers and the general monthly journals aided in portraying this image by giving sporadic attention to the progress of wireless during the 1910s. Disillusioned by the recent years of scandalous wireless companies, the print media made it easy to believe that radio was an obscure technology with little bearing on everyday life.

But that is not to say that the average Houstonian did not get exposed to wireless. During the 1910s, wireless was occasionally shown in public exhibits and demonstrations. The United States Navy visited Houston in November 1912 to show how a wireless station functioned at the city's annual No-Tsu-Oh carnival.[6] With heightened emphasis placed on the efforts to secure an inland seaport for Houston, the festival featured more maritime related presentations, including demonstrations on how the Navy transmitted radio messages throughout her fleet. For the demonstration, Navy representatives installed temporary, small-scaled models of radio stations on the festival's floor.

Houston's first academic institution also took an interest in wireless at this time. The Rice Institute, changed to Rice University in 1960, opened in 1912 with a complete physics department headed by nationally renowned physicist Professor Harold Albert Wilson. He conducted experiments with making radio waves on campus, and by 1915 Professor Wilson gave a lecture and demonstration entitled, "Electric Waves and Wireless Telegraphy."[7] Covered by the *Houston Chronicle*, the lecture detailed the physics of sending and receiving radio signals.

> "A rapid electrical oscillation in a conductor produces a series of electric waves which travel outward from the conductor into the surrounding space. These waves can be detected and their properties examined by means of a coherer. Such electric waves are of the same nature as light and travel through empty space with the velocity of light, or 186,000 miles per second. In wireless telegraphy, an

> electrical oscillation is produced in a conductor that extends upwards from the ground to a considerable height. When these waves fall on another similar conductor, which may be many hundreds of miles from the first one, they produce an electrical oscillation in it which can be detected by means of a coherer or other suitable device."[8]

What was described that day and shown in a laboratory environment with electricity radiating from the equipment were the basic principles of wireless that for years had been practiced by amateurs. The electrical oscillation described in the experiment was the generation of radio waves from the sending apparatus. The coherer described is the actual radio receiver.

Without a doubt, an important factor for the spreading popularity of radio among young boys was the emergence of the wireless clubs. A subculture comprised of dedicated enthusiasts, these wireless clubs first began to appear around urban areas of the nation in 1909, and by the early 1910s every large city had at least one. Through these clubs, amateurs shared technical information and informal fraternal bonds among these young men and boys strengthened. More importantly, the wireless club gave the amateur an added lobbying instrument against persons and organizations that were pressing down on them and attempting to restrict their activities. Even after the Radio Act of 1912, talk still continued to ban the amateur entirely from the airwaves. Eventually, this vast network of wireless clubs scattered throughout the nation led to the formation of the American Radio Relay League which provided the much-needed advocate voice for the amateur who for years had been the target of much scrutiny in government circles.

Houston's first large wireless club may have been the Houston Amateur Wireless Association. It was formed in the early 1910s with meetings held weekly in the newly created YMCA building in downtown. The agenda called for the technical assimilation of wireless telegraphy among its members and anyone else interested. James Bowling served as president of this club in 1916.[9] Aside from this, Bowling also holds the distinction as being the first Houstonian to obtain a license for his amateur station in 1915. His career was in railroad, with his first employer being the Sunset Central Lines. He later went to the Southern Pacific Railroad Company in 1917. As a railroad station agent, he was trained as a telegrapher, an occupation that many other Houston amateurs seemed to have pursued. As a hobby, Bowling experimented in wireless, upgrading the strength of the transformer on his station, designated with the call letters 5CO, to a half-kilowatt in 1916.[10]

By 1916, years of practice made some Houston amateurs quite accomplished in their hobby, and a few may have started dabbling in the next development of wireless, the use of radiotelephony as a means to transmit. Not fully realizing what their actions were leading to, these amateurs were laying the groundwork for a whole new dimension of radio.

At this point, a closer examination of some of the more important amateurs before 1917 and how they helped shape early Houston radio is in order. Surely one of the most influential wireless operators at this time was Louis Peine. Born around 1900, Peine was the son of Wolly and Nettie Peine of Hungerford, Texas. The first of five children, Louis Peine was born in Flatonia, Texas where his father served as a station agent on the Southern Pacific railroad line that ran into Mexico and was laid out roughly along the Gulf Coast of Texas. As a station agent, Wolly Peine served as a telegrapher, receiving and sending wire messages pertaining to the trains that ran through Flatonia. Shortly after Louis Peine was born, the family moved to Houston where Wolly Peine was employed as an insurance agent for a local company. It was not long after this that he accepted a position as the business manager for the Houston School System, later the Houston Independent School District.[11]

The Peine family eventually settled on Rosalie Street, an upper middle-class area southwest of downtown where young Louis Peine had many friends to play with. This area bred the city's largest concentration of amateur operators and quickly Louis gravitated to wireless and became an accomplished amateur operator. His home life may have also fueled his interest in wireless. Family accounts note that Wolly Peine passed on his knowledge of Morse code communication to his son.[12]

Another amateur important to Houston radio at this time was James Lockhart Autry III. Affectionately known as Jimmie,[13] the young Autry was a fortunate son from a wealthy family. Born in 1899, Jimmie came from Corsicana, a small town in central Texas, which became a mecca for the state's budding oil industry before 1900. Jimmie's father, James Lockhart Autry II, had come to Corsicana to claim land awarded to his grandfather, Micajah Autry, a veteran of the Alamo battle. James Autry II was educated as an attorney and specialized in petroleum law, coming into the service of Joseph Cullinan, the Texas Company's founder. Promptly, Cullinan appointed Autry as the company's attorney general and as one of its first directors.[14]

Major oil strikes in the area of Spindletop near Beaumont in 1901 changed Corsicana's status as the state's oil capitol. Hordes of wildcatters converged on the Beaumont area to take their chances on the booming oil industry, including the Texas Company. The first of two children, Jimmie Autry moved with his family to Beaumont in 1902 where the Texas Company established its headquarters. After a few years, the company once again moved its offices to Houston and along followed the Autry family.

Once in Houston, the Autry family eventually resided in Courtlandt Place, a newly created exclusive strip of wealthy homes with some of the city's "new money" brought in during the oil boom. There, Jimmie Autry got involved in wireless. In military school, he reached the rank of officer and was already developing an interest in long distance communication when he joined his school's Signal Corps,[15] a discipline most likely training in the art of telegraphy. At this time, his father's company, the Texas Company, installed an impressive wireless system along the upper gulf coast that may have also energized Autry's interest in wireless. By the age of fifteen, Jimmie had become an expert amateur with the most powerful and state of the art equipment at his hands.

The Autry family would take long summer excursions from Houston, allowing Jimmie to gain wireless experience he would not otherwise get had he stayed in town. In 1915, he worked as an assistant wireless operator on a Great Lakes line of steamships. That same summer he went to Chicago where he passed the examination to become licensed as a second-class operator. The following year, he secured a first-class license for his station with the call letters 5ED.[16] It was the second such license and the last obtained in Houston before World War I. His powerful amateur station was able to operate to the maximum allowed at one kilowatt. By late 1916, station 5ED was able to send out messages at a radius of 1,100 miles and, during the winter of 1916-17, Autry was heard in 22 states around the nation.[17] A family summer trip to California in 1917 allowed Jimmie to function as a wireless operator on a Pacific Ocean steamship.[18] As a member of the American Radio Relay League, Jimmie was responsibility for maintaining his station in connection with others that served on a wireless trunk relay line from Chicago, going through the Midwest and ending in Texas. During times of emergency, Autry needed to have his station prepared to relay any radio messages that passed through his district.

Even though he was not as wealthy as Jimmie Autry was, Clifford Wise Vick rivaled the best of Houston's amateurs. A graduate of the Heights High School, Clifford Vick was a Yoakum, Texas native, born in 1896.[19] His father, Ransom Vick, was employed in the railroad industry and later moved his family to Houston. After holding various jobs, Ransom Vick eventually landed a position managing an automobile service station. Clifford Vick was one of the earliest Houston amateurs, first getting involved in wireless by building receiving sets in 1908 and eavesdropping on transmitting radio traffic. This lasted a few years until 1911 when he decided to begin sending messages. Before the United States' entrance into World War I in 1917, Vick was reaching other operators in San Francisco and New York. With his best receiving set, Vick was able to retrieve signals from as far as the station transmitting from the Eiffel Tower in Paris, France. Another of his successes was reaching an important military station in Nauen, Germany,[20] where before and during the war, much wireless traffic was transmitted.

Louis Peine, Jimmie Autry and Clifford Vick became close friends through their common interest in wireless. They, along with a few other young men, formed another club, more exclusive and private than the Houston Amateur Wireless Association convening at the YMCA. Called the Houston Radio Club and formed approximately in early 1916,[21] this group was comprised of the best Houston amateurs. New members were accepted by invitation only. Susan Douglas, in her book *Inventing American Broadcasting*, points out that these clubs often held their meetings on the air with a prearranged frequency and each members would acknowledge by signing on and transmitting his license's call letters.[22] Other members of the Houston Radio Club were native Houstonian Alfred Neal Dargan, the club's original president, Felix Currie and James William Weatherford,[23] the son of a wealthy cotton firm owner in town. Humorous stories tell of James Weatherford's proficiency in electrical applications. Brown-outs and temporary blackouts would occur in the Weatherford household when the young James Weatherford would initially rev up his wireless set to begin sending and receiving messages. As a deterrent to unwanted visitors to his room, he would lace the stairs and doorways leading to his room with wires connected to a live, small storage battery.[24] Unexpecting victims would often receive a small electrically generated jolt as a warning not to enter.

In a short period of time, the Houston Radio Club became the leading wireless group in the city. In November 1916, the club held a wireless exhibition at the city's popular "Electric Week," seven days of a carnival atmosphere which included a street parade in which local vendors and groups showcased the wonders of electricity through the demonstrations of appliances and experimentations. For the showcase convention, the Houston Radio Club installed a temporary wireless set in the city auditorium. Gathering crowds watched as members of the elite club received time signals from a government-sponsored station in Arlington, Virginia,[25] a favorite at the time for amateurs to listen in on. During the course of the week, the Houston Radio Club station may have received other interesting transmissions from Arlington as the station was also conducting periodic experiments with sending out messages through radiotelephony by this time.

Amateurs, for the most part, still relied on their inventiveness to build sets. Makeshift radios continued to be common and a few businesses now began to cater to these innovative Houston amateurs. One of these business owners was Ben James Still. As part owner of a firm of electrical contractors in Houston, Still envisioned a new venture for his business as many boys would come to purchase electrical parts for their radio sets. Ben Still moved to Houston from Mississippi in 1901 to start a business with his brother, William Still, and his brother-in-law, Alfred Hurlburt. Called the Hurlburt-Still Electric Company, the firm began as electrical contractors installing electrical systems in homes and businesses.

Later, the company diversified into a store selling items such as electrical appliances and added a service garage equipped to servicing electrical systems in automobiles.[26] Family legends of the Still family maintain that members of the Hurlburt-Still company were some of the first in the city to use Christmas lights on their nativity tree, attracting the attention of many neighbors every Christmas season.[27] Most likely the idea to open a wireless section in the company came from Ben Still. All evidence shows that he took the most active interest in wireless in the company. So equipped was the Hurlburt-Still Electric Company to serve wireless enthusiasts that radio operators on ships would travel from Galveston to purchase equipment.[28]

From 1913 to 1917 historians have pointed out that the increase in the number of amateurs also brought an increase in some of them experimenting with voice transmissions. Most amateurs were still sending messages in the form of dots and dashes, though a few more fortunate ones were beginning to use Lee de Forest's invention of the audion. One of these may have been Jimmie Autry who seems to have been using de Forest's prized audion to receive and send radiotelephonic transmissions during the 1910s. In 1915, the American Telephone & Telegraph Company (AT&T), as commissioned by the United States Navy, experimented with wireless and successfully installed a transcontinental radiotelephone service that functioned from the naval installation in Arlington, Virginia to as far away as the Hawaiian Islands. Soon, amateurs became frequent listeners to these stations' experiments to transmit wireless telephony.[29] Voice messages and music were aired on these transcontinental transmissions on a regular basis and some Houston amateurs were fortunate to hear them. Houston Radio Club member Clifford Vick frequently succeeded in listening in on the Hawaiian station's transmissions of music, so well in fact that "he could almost see hula girls dancing."[30]

Closer to Houston than the stations in Virginia and Hawaii, military campaigns in Texas were also experimenting with primitive forms of broadcasting prior to the United States' involvement in the First World War. For the 1916 United States presidential election, an army installation known as Fort Sam Houston near San Antonio transmitted the election's returns to other army bases and camps situated along the Texas-Mexico border during United States General John J. Pershing's massive failed expedition to capture Mexican revolutionist Francisco "Pancho" Villa.

Armed with cutting-edge technological inventions such as a compact armored vehicle and the airplane, General Pershing also incorporated radio communication to notify his troops of the outcome in the Woodrow Wilson-Charles Evans Hughes presidential race. On November 7, radio operators at Fort Sam Houston began tapping out on Morse code the elections' returns informing all listening that Woodrow Wilson had won a second term with a narrow margin. Among those receiving this isolated "broadcast" of election returns may have been members of the Houston Radio Club, including Jimmie Autry who was a frequent listener to Fort Sam Houston.

Another Houston character who deserves some attention in our history is Howard Robard Hughes II. Before becoming a famous billionaire, a film producer in Hollywood, a record-breaking aviator, and a legendary recluse during his last years, young Howard Hughes was a radio enthusiast in Houston. An only child from a privileged oil family, Hughes' early fascination with mechanics led him to radio communication. Encouraged by his father, he enjoyed tinkering in the family workshop where he built his radio sets. He was also fortunate enough to live in the area of town where many of the best amateurs resided.

Living in close proximity to some members of the Houston Radio Club made it convenient for Hughes to learn from the best. Louis Peine's younger brother, Leon Peine, recalled how a young, quiet Howard Hughes would often visit to learn the principles of wireless. "Hughes would come on his bike to visit my brother, Louis, and

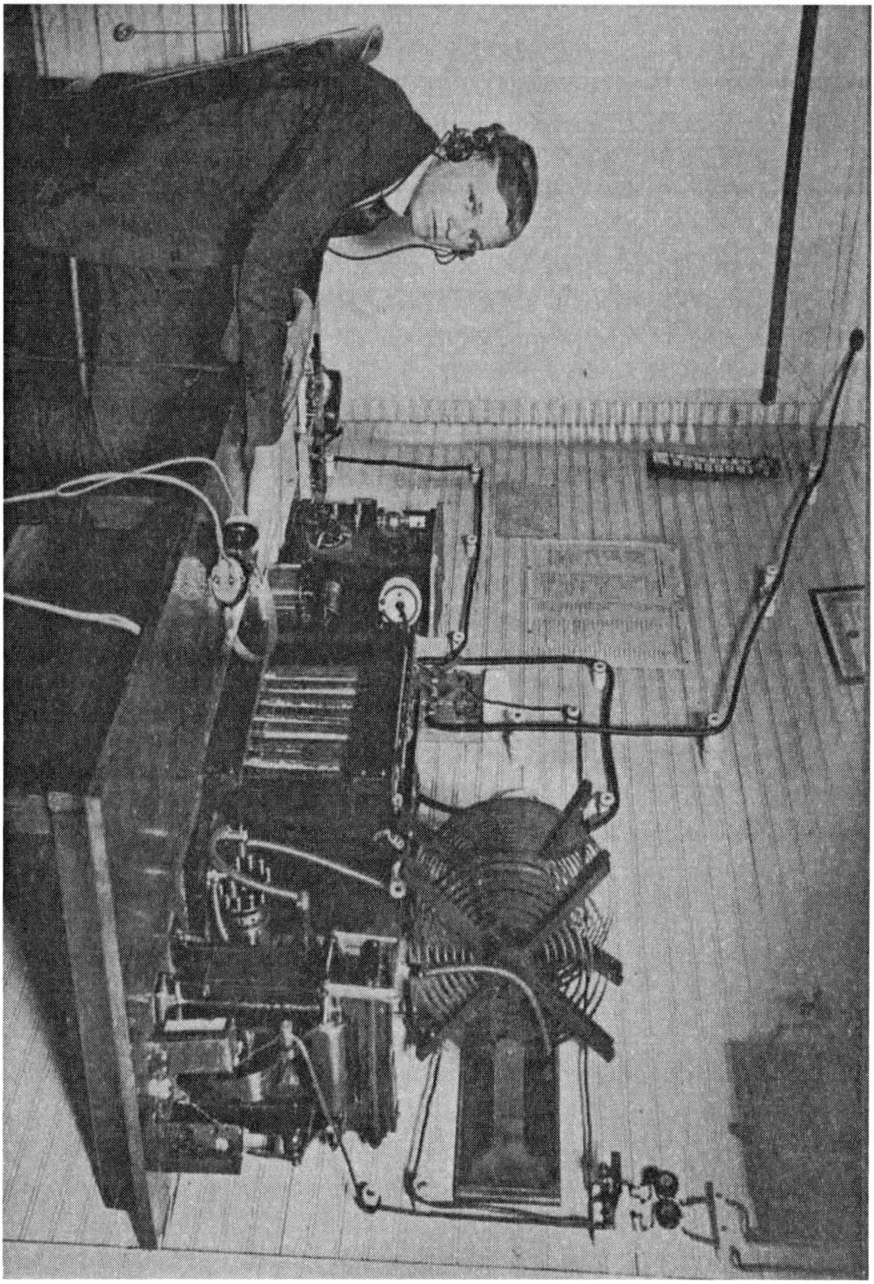

Jimmie Autry and his wireless apparatus, circa 1916. *Courtesy of the Woodson Research Center, Rice University*

would spend hours on end with him in our backyard learning how to construct radios."[31] Actually, Hughes lived closer to and ran in the same social circles as Jimmie Autry, certainly another of his early wireless mentors. A close friend of Hughes said that he learned Morse code in one day and immediately began flashing messages to ships out in the Gulf of Mexico. Hughes was also quoted as saying that his first makeshift radio was "the proudest thing he ever made." A biography on him claims that among his peers he was the best and had the finest equipment, enough for him to form a club called the Radio Relay League.[32]

Up to 1917 Houston amateurs had survived government regulation and the controversial side of wireless brought by the scandalous companies exploiting it. The majority of amateurs continued their activities with little impact to the average citizen. But that would change in a few years. Radio began in the early 1900s with a minimum number of operators conversing through dots and dashes over simple crystal sets. Now in the late 1910s, radio communication had evolved to include more improved equipment. Wireless telegraphy was slowly being supplanted by radiotelephony as more amateurs and organizations started exploring voice and music transmissions. A new concept whereby a station would transmit a message to many, scattered stations was also being invented at this time. Today we know this concept as a "broadcast," a term unheard of before 1921.

But all of that would be interrupted in April 1917. The Great War that had been raging in Europe could no longer escape the attentions of the United States. This monumental event would transform the definition of radio. Wireless would no longer be exclusively a young man's activity. Instead, it was changed into something that would concern the whole country and become interwoven in the fabric of American life.

Chapter Five

Jimmie Autry shot his pistol into the air on the morning of April 6, 1917 to tell his neighbors in Courtlandt Place that the United States had officially entered the First World War.[1] He may have been one of the first Houstonians to know of the event when he heard the news over his wireless set. It would also be the last thing he would be permitted to hear via wireless for the next eighteen months. A clause in the Radio Act of 1912 allowed the federal government to restrict the use of radio waves during times of distress.[2] Immediately the clause was enacted, and all radio use that was not authorized by the federal government was prohibited.

The years enjoyed by amateurs when radio laws were not strictly enforced had come to an end. While it would seem like an opportune time for mischievous amateurs to eavesdrop secretly on classified government messages during wartime, the vast majority of them complied with the radio ban. Some went as far as dismantling and storing their stations while others merely discontinued the activity and never resumed it following the war. Entities using wireless in their commercial activities were also ordered to shut down unless their use was deemed necessary to continue operating. In such cases use was controlled, maintained and operated by the federal government.

The government was adamant about enforcing the ban on radio and was swift to investigate any possible violations. Shortly after the declaration of war, reports had reached local authorities that Jimmie Autry may not have completely disabled his station. A private, active station was perceived as a serious threat to the country's security. Already recognized around the city as one of the best amateurs, Autry could have been charged with treason by violating the wireless ban.

Prior to the war, Autry had conveniently used the antenna's aerial in his yard as a flagpole. During the ban, he apparently disabled the aerial by removing the wireless paraphernalia. The Autry family, though, decided to continue using the aerial as a flagpole. Standing 81 feet high, the aerial, and now flag mast,[3] could easily be seen in much of the immediate vicinity. Uneasy with the decision, local law enforcement agencies wanted the complete dismantling of the radio station, including the mast. Jimmie's assurances during an initial interview that his wireless ability was disabled were apparently not enough for some authorities. A second visit to the Autry residence was deemed warranted with the intention of having the pole taken down. Local FBI agent W. W. Green and the Houston chief of police tried to arrange a second meeting in the hopes of resolving what could have easily exploded into a serious matter. Unable to meet with them at this second meeting, James Autry II assured the officials that his son's aerial would not be used for transmitting or receiving messages[4] and records remain unclear as to whether it was allowed to stand.

As Autry's situation showed, the United States government during the outbreak of the war viewed the amateur as a potential problem. For years, the government had heard stories of how amateurs were nuisances and measures were taken to restrict their activities. However, radio was a powerful weapon in combat and the United States discovered it had a desperate shortage of enlisted men without wireless skills. Suddenly, the amateur became an important component in the military equation.

Badly in need of enlistees who had radio experience, the government turned to the amateur for aid. Some Houston amateurs went abroad to do "their part" while others remained in the city's home front to serve as wireless operators. Most of those who would later become pioneers in Houston's radio broadcasting craze of the 1920s were a little too young to serve in World War I's front lines. Instead, they temporarily relieved adult wireless operators on civilian merchant ships called to serve in the United States military. Aside from Jimmie Autry, who acted as a wireless operator on a Pacific Ocean steamship in the summer of 1917,[5] several local amateurs migrated during the summer months to the Great Lakes area where the maritime industry desperately needed wireless operators. First president of the Houston Radio Club, Alfred Neal Dargan, heard the call and served in the newly created Naval Reserve in the Great Lakes area where he gained valuable experience as a wireless instructor and operator.[6]

Jimmie Autry's imposing antenna for his station 5ED caused a stir with local authorities when he decided to fly an American flag from it when the United States officially declared war against Germany in 1917. *Courtesy of the Woodson Research Center, Rice University*

Many new enlistees joining the war effort with little or no wireless experience made the federal government realize it needed to start a program to train soldiers in wireless technology. Cooperating with the federal government, the Houston School System implemented such a program in the fall of 1917[7] that survived for years following the war. Initiated by Wolly Peine, an evening course was held on the school grounds of Houston's Central High School beginning in November. Instruction at an accelerated rate was held five nights a week from

7:30 to 9:30 p.m. Applicants for the class were screened based on their previous experience in telegraphy-wire or wireless and the passing of other aptitude tests. An enlistee with such training had a far better chance of entering the service as a low-ranking officer rather than as a private.[8]

As a regular administrator in the Houston School System, Wolly Peine made an excellent candidate to instruct this class. His familiarity with the school system and his previous experience as a telegrapher for the railroad industry gave him the opportunity to serve his country. But Wolly Peine offered more in the fact that his son Louis Peine was also a seasoned wireless operator, ready to instruct a course quickly overcrowding with students for only one class. Soon, a second evening wireless course began, instructed by Louis Peine. As explained earlier in this history, Wolly Peine taught his son telegraphy and Morse code. Now the circle was complete as father and son offered their wireless expertise and contributed to the war effort on Houston's front lines.[9]

The number of Houston enlistees eager to learn wireless continued to grow. The Signal Corps needed more wireless operators, and fast. Agreeing to pay half of the instructors' salaries, the federal government increased the successful program's budget and a day class was added in December 1917. Still held at Central High School, the class was taught by Louis' fellow member of the Houston Radio Club, Clifford Vick.[10] It can be said that even though the instruction dealt with wireless communication, emphasis seems to have been placed on educating students in telegraphy and the radio codes used by the Army.

The Houston School System was not the only higher education institution in town offering wireless courses for enlistees. Similar classes were being held at the Rice Institute. Professor of electrical engineering at Houston's only university, Nicholas Diamant, had developed an early interest in wireless before the war. A member of the institute's faculty for several years, Professor Diamant agreed to instruct radio on campus during wartime. The U. S. Signal Corps provided an apparatus called the omnigraph with which students could practice sending and receiving messages without requiring the availability of an entire radio set.[11]

Houston amateurs helped in other unique ways, too. The scarcity of government wireless stations in the area was quite evident when authorities commissioned the assistance of experienced radio operators in the sending and receiving of classified messages. Family accounts recall that James William Weatherford, an original member of the Houston Radio Club, was called upon to act as a government-sponsored station. Military officials occasionally visited the Weatherford residence to use James Weatherford's services as a wireless operator. The receiving and sending of messages were classified in nature and even close family members were forbidden to learn the nature of the transmissions.[12]

After the United States' entrance into the war, military bases began springing up around the nation, and efforts were underway by local officials to secure some of them in Houston. The first of these installations was Camp Logan, a National Guard training facility just several miles west of central Houston. By the summer of 1917, the Houston area had also secured the construction of an air base. Named after deceased aviator Eric Ellington, the aviation installation was located about thirteen miles south from downtown Houston near the town of Pasadena. Work was soon completed in late 1917 on Ellington Field and the prompt arrival of bi-winged airplanes, equipment and pilots began arriving in November. Considered an advanced training facility for pilots, Ellington Field taught everything from dropping live bombs into simulated battlefields to instruction on how to train homing pigeons to carry messages.[13]

Of course, this new technology called radio communication was quickly viewed as a new dimension to be added to military aviation. Even before the notion that airplanes could be used in combat by dropping bombs or spraying the enemy with mortar from the air, military aviation was viewed as one's "eye in the sky." Aviators were often sent to bring back reports on enemy troop movements and the placement of enemy arsenal and batteries. Also, during battle, an aircraft used as an observer could direct the fire of its allies' batteries. Learned in wireless communication and equipped with a radio on board, the reconnaissance aviator could now send immediate reports and receive orders from the ground while still in flight. Quite obviously, the Army took a keen interest in radio and was prepared to seriously experiment with it in connection to military aviation.

As some of the first biplanes were leaving the runways on practice maneuvers from Ellington in January 1918, a radio department was established on the air base to train pilots in air-to-ground communication and experiment with radio communication. The radio department initially was comprised of one officer, seven cadets and two civilian radio engineers.[14] One of these civilian engineers was Jimmie Autry. Apparently drawing wide attention for his ability in wireless, Autry was given the privilege of aiding in the installation of radio equipment on the base. For months, he was given full access to the base and conducted wireless demonstrations as communications were established between planes in flight and the ground base. Dividing his time between Ellington Field and his duties as a corporal for the S. A. T. C. (Student Army Training Corps) for the Rice Institute, Autry was given the honorary position of consultant radio engineer for the air base.[15]

Early in 1918, and seeing an opportunity to serve his country and further his interest in wireless, Alfred Daniel decided to enlist in the war as a radio technician. For two reasons Daniel may have viewed this moment as a perfect time to join the Army. Ellington Field was the closest military facility that offered the most modern training in radio communication. A tour of duty at the base's radio school also included an initial training period in the Air Service Radio School at the University of Texas in Daniel's birthplace of Austin, Texas. The Air Service Radio School at the state's University was said to be one of the first of its kind in the country,[16] an attractive item for any wireless operator wishing to learn this new dimension of radio communication in connection with military aviation.

Daniel's experience of more than ten years gave him the necessary leverage to have his preference granted to serve at Ellington Field. Daniel's service began on April 18, 1918 when he was 28 years old, and for the next eleven months he would be attached to the Army learning and applying the cutting-edge technology of radio in relation to aviation.[17] For the first two months of his service, he was stationed at the military radio school at the University of Texas. Upon his return to Ellington Field, his primary duty was to instruct service men in the science of radio at a time when the average soldier had little understanding of the basic principles in wireless. The aircraft's radio equipment he had to work with was often simply an upgraded version of the crystal set connected to an antenna comprised of a sensitized wire dangling midair behind the aircraft in flight after the aviator unraveled it from the cockpit. One of Daniel's classes included a basic tuning practice session in which aviator trainees would be graded on how quickly they could tune into a station in a specified time period. A student's performance in this session was also measured by how much radio interference he eliminated as the station was tuned. Eventually, the grounded class would lead the student into a flying aircraft where the tuning process proved to be more challenging.[18]

With the rank of Private and a title of "unassigned," Daniel was placed with all other radiomen at Ellington in squadron C. Life on the air base must have somewhat pleasant for enlistees, and especially for Alfred Daniel, as they were often given weekend passes into Houston, which included transportation by a railroad line that ran beside the military base and into the heart of the city.

Aside from keening his knowledge in radio communication, military life also allowed Private Daniel to perfect his talent as cartoon illustrator. Some of his drawings parodying radio and aviation men appeared in Ellington Field's first (and only) yearbook in 1918.[19]

Months into Alfred Daniel's tour, the radio school at Ellington made a decision to advance and involve itself in the experimentation and training of radiotelephony. Since 1915, the Allies forces in Europe had been conducting small, isolated experiments to direct flying squadrons through radio voice commands.[20] For aviation, wireless telegraphy had quickly become a bothersome and obsolete form of communication. Air bases needed gradually to concentrate on the perfecting of radiotelephony for aviation purposes, and Ellington Field needed to keep up with the times.

By the summer of 1918, Ellington Field's radio school had started to equip itself with a radiotelephony facility. Once again, Jimmie Autry may have served as consultant radio engineer as the military installation upscaled some of its ground facilities and aircrafts with voice transmitting equipment. The air base's radio facility created another division called the Advanced Radio School, chiefly dedicated to the advancement of radio transmitted voice communications. Private Daniel proved to be a fast learner and skilled radio technician when he was quickly promoted into the Advance Radio School where he had access to the experiments the air base was conducting with radiotelephony. No official, detailed records exist concerning the radiotelephony tests at Ellington Field. However, informal records state that tests on the base were successful in establishing voice communications between an aircraft in flight and a grounded facility, reaching a distance of several miles between the two parties.[21]

In November 1918, Daniel and his fellow instructors in the Advanced Radio School were assigned to install a radio school at a new air base just south of Austin, Texas, called Penn Field. Like at Ellington, his assignment was to establish a radio facility and instruct aviators in the science of wireless. However, just two weeks into the assignment Private Daniel learned that the armistice had been signed and World War I had come to a conclusion.

World War I service men at Ellington Field learning the basics of radio communication. *Courtesy of Houston Metropolitan Research Center*

Underground bunkers used as radio shacks at Ellington Field in 1918. *Courtesy of Houston Metropolitan Research Center*

Private Daniel would immediately return to his home air base where he would serve his remaining tour until March of 1919.[22]

With the termination of the Great War, wireless amateurs gained some respect in the public eye. Houston amateurs, in their unique way, helped the cause and can be honorably mentioned for doing their part to win the war. However, the effects of wartime vigilance had not yet subsided. The federal government still maintained control of the airwaves and soon bureaucratic talk resumed to maintain the airwaves off limits to the private sector. Fortunately, the nationwide advocate group for amateur operators called the American Radio Relay League successfully lobbied the federal government to relinquish the airwaves for public use. Removing the ban, however, would be slow and gradual, not to take place until April 17, 1919 when amateurs were again allowed to hear and monitor messages.[23] Continuing with wartime vigilance, the federal government remained fearful that amateurs could send out false or confusing messages and interfere with radio traffic. The privilege to transmit messages of a private natural was still forbidden.

The federal government's act to allow amateurs again to monitor and receive radio messages was perceived as a step in the right direction, though. Consequently, amateurs began to reorganize radio clubs that once convened before World War I. In Houston, founding members of the Houston Radio Club began holding informal meetings at South End Junior High School[24] in preparation for holding a large recruitment drive. It seems that no other large radio club existed to compete with them. The Houston Amateur Wireless Association that met weekly at the YMCA apparently did not survive the war.

Public announcements appeared in local newspapers asking for all persons interested in wireless to join the "reorganized" Houston Radio Club.[25] In September 1919 at the initial meeting held at the downtown YMCA, dozens of new amateurs joined. Returning as club president was Alfred Neal Dargan with

Jimmie Autry serving as vice-president. Louis Peine acted as secretary and James William Weatherford served as treasurer. Among the new members were Hallet Worthington and Ingham S. Roberts III. Worthington had presided over a radio club in Aberdeen, North Dakota, and later served in the Naval Reserve on the Great Lakes during World War I. Ingham Roberts III was a third generation Houstonian and son of a prominent attorney and noted Texas historian. Aside from the opportunity for amateurs to meet and exchange wireless issues, the Houston Radio Club maintained the evening course in radio instruction that had been started during World War I by the federal government and the Houston School System. By November, the Houston Radio Club became a nationally recognized body when it was admitted as a member of the American Radio Relay League with Hallet Worthington appointed as the club's city manager for the League.[26]

Clifford Vick was monitoring the airwaves hoping to receive any information on an unnamed hurricane that was lingering off the gulf coast near Corpus Christi. Due to the government's ban on sending messages still in effect, Vick could only listen as a message was being sent from the Texas Company's steamship *New York* while out in the Gulf of Mexico in the early morning hours of September 14, 1919:

"Latitude 20.18, longitude 92.39, barometer 29.69, winds 38 to 40 miles, northeast and increasing for the past hour."[27]

An hour later, at 2 a.m., the *New York* reported an update:

"Barometer falling. Wind blowing with increased force."[28]

The hurricane was expected to hit between Corpus Christi and Brownsville, Texas sometime later that day. Realizing that he had crucial information, Vick telephoned the *Houston Chronicle*. Knowing that more news would be reported, Vick returned to his radio. The storm did in fact strike the Texas coast sometime later that day with winds at 110 miles per hour while Vick continued his vigilance over his radio in the hopes of receiving dispatches from the storm struck area. The *Houston Chronicle* reported on September 15 that Clifford Vick received the first messages from Port Aransas and Corpus Christi from an U. S. customs agent stating that the area was severely damaged by the storm, including damage to land communication lines. Other messages received by the Houston amateur included a distress signal from the steamship *Median* saying that she was run ashore during the hurricane.[29]

Similar to another incident involving the *Titanic* disaster of 1912 when amateur David Sarnoff was providing near exclusive information, Clifford Vick was proclaimed a local hero. To his credit, he did not dwell in the limelight and may have actually avoided the attention he was receiving. In reality, Vick was not the only operator picking up messages during the hurricane disaster as was reported in the *Houston Chronicle*. A telephone company station equipped with a radio in Galveston was also receiving wireless messages.[30] Furthermore, it would also be logical to believe that other amateurs in the Houston Radio Club may have also participated in monitoring the airwaves during the storm.[31] Nevertheless, the *Houston Chronicle* had their "hero," and the incident was reported as such. A week following the storm, the *Houston Chronicle* printed an interview with Vick. In the article, he lamented how because of the wartime ban he was unable to send messages. Vick commented, "Think what might have happened if I could have had a sending apparatus during the storm. I could have asked questions and probably obtained more information than I did."[32]

How fate seems wicked at times. A week later, on September 26, the federal government announced the removal of the ban. While a strickened lower Texas gulf coast area, with $20 million in damages and 284 lives lost, sought to rebuild itself, jubilant screams from members of the Houston Radio Club erupted at their late September meeting after hearing the news that amateurs had regained the privilege to transmit messages.[33]

History can debate what impact World War I had on the evolution of radio communication. An argument can be made that the war hampered what progressions American radio had made before 1917. As we have seen, there was the development of radiotelephony and the idea of broadcasting was already on the minds of some. The war interrupted the efforts of these few inventive amateurs and scientists. But radiotelephony advances made by governmental entities in World War I may have actually accelerated events leading to the development of mass broadcasting in a post-War environment.

Lee de Forest's audion, modified and improved, was now called the vacuum tube and during the war years the federal government mass-produced it for many of its radio needs. Unlike the sparsely distributed audion, the vacuum tube was overproduced and subsequently, a huge distribution spilled over into post-war America that would set off a chain of events in radio broadcasting history. Many service men returning from the war now had access and were trained in the vacuum tube, the key essential element for radiotelephony.

Houston amateurs started seeing the impact of the vacuum tube by August of 1919. *QST*, a monthly publication from the American Radio Relay League, reported news coming from Houston in August that Donald Harper Graham from a small rural community just southwest of Houston was attempting to establish contact with the city using vacuum tubes. Incidentally, Graham was already trained in radiotelephony when he served in World War I as a radio technician with Alfred Daniel at Ellington Field. Graham lived in House, Texas, a small farming area twenty miles from Houston in Fort Bend County. Living several miles from telephone and wire telegraph communications, Graham intended to transmit voice messages to a Houston station to order supplies for his farm. The *QST* report did not report to whom Donald Graham wanted to send voice messages to.[34] But evidence indicates that at least several people in Houston were also experimenting regularly with radiotelephony by this time.

When Clifford Vick became a local household name from his reports concerning the hurricane of 1919, one of the things he described on his radio set when interviewed by the *Houston Chronicle* was the "audion detector,"[35] de Forest's old audion now referred to as the vacuum tube. As mentioned in Chapter 4, Vick may have already been experimenting with radiotelephony for quite some time before World War I when he would eavesdrop on the federal government's transmissions of music and voice messages from government stations in Virginia and Hawaii. Of course, Donald Graham's fellow serviceman Alfred Daniel may have also been the intended recipient of Graham voice messages from his farm. Indeed, another likely intended Houston station that Donald Graham wanted to reach was that of the Hurlburt-Still Electric Company, the area's leading supplier of wireless items, including vacuum tubes.

Mrs. Lorene Kaiser remembered the excitement in post-World War I Houston as amateur operators were returning on the air to produce voice transmissions. Her older brother was Louis Peine, veteran of Houston's airwaves. One evening, as a young girl, Mrs. Kaiser remembers seeing her brother hurrying around the house excitedly. With no else to share in his excitement, Louis grabbed his sister's hand and took her to his radio shed in the backyard. To her astonishment, Louis proudly showed off the radio set he had barely equipped with vacuum tubes and was now receiving a clamor of different voices and conversations over the airwaves.[36] Houston was back on the air with a fury.

Howard Hughes Sr. was facing a dilemma that many parents encounter when they try to give their children the best possible advantages in life. How do you get your child interested in pursuing studies in a prestigious institution? A proven good study with his hands, Howard Hughes Jr. made own his radio sets and constructed a motorized bicycle but was a mediocre student at best who was uninterested in academics. Perhaps to win status for his son or to incorporate the young Hughes' interest in wireless in a university environment, the elder Hughes donated a powerful radio set to the Rice Institute in early 1920.[37] Although it would be another three years before Hughes Jr. would attempt college (including a brief semester at Rice), the donated radio set provided the seed to start the university's involvement in radio. Up to this point, the radio scene at the Rice Institute was in a neglected condition since Professor Diamant's war courses in wireless. Other colleges around the state such as Texas A&M University still had apparatus maintained since the days that John Dickinson of the Texas Fiscal Agency had donated the equipment in the early 1910s. The Rice Institute felt a need to maintain status with other Texas universities. With Hughes' donation, the university was granted a radio license with the call letters 5YI. Station 5YI was housed in the electrical engineering lab on campus and was maintained by students and faculty. No formal wireless courses were offered in relation to the Rice station but anyone associated with the college was given the access to use it. Although general communication was still welcomed, radio activities on campus became organized in the fall semester of 1920 with a radio club and the members emphasized establishing wireless communication links with other colleges around the nation, especially with those in Texas.[38]

Student of the Rice Institute and member of the Houston Radio Club, Athna Ellis was elected president for the college's first radio club, the Rice Radio Club, in September of 1920. Assisting Ellis was E. F. Carter, fellow student, and M. K. Humphrey head of the electrical engineering department. The immediate agenda for the Rice Radio Club called to establish a daily wireless exchange with the state's other major universities equipped with radio stations. The University of Texas and Texas A&M University agreed to maintain daily radio communication with the Rice Institute. Of the three colleges, Texas A&M was the best equipped and trained to handle wireless messages at this time. Even though the University of Texas had agreed to participate in the three-college wireless link, it had no organized wireless group. Instead, it relied on amateurs in Austin to send its messages.[39]

In order for the three colleges to receive and send messages with no radio interference from other local stations transmitting, the Houston Radio Club members abstained and asked others to remain off the air everyday from 5:00 to 5:30 p.m. This half-hour allowed Rice amateur operators to exchange campus news and announcements with the other Texas colleges.[40] Radio communication became so popular on campus that plans were made to have members of the Rice Radio Club rotate shifts on station 5YI to constantly monitor all wireless traffic during the daylight and early evening hours.

These humble beginnings on Texas' campuses present a perfect example of how broadcasting would evolve from point-to-point radio communication. As early as September 1920, efforts were underway to have the colleges broadcast final results of athletic events immediately at the games' finales. A simple progression from this would be to broadcast the event as it occurred. Later, in this history, it will be discussed how this radio link between Texas universities gestated into an event credited as the state's first broadcast of a live sports event.

Elsewhere in Houston, amateurs had broadcasting on their minds. Long time friends Louis Peine and Alfred Neil Dargan decided to construct a wireless station above the Carter building in late 1919, the exact place where the Texas Wireless Telegraph & Telephone Company installed its station in the early 1910s. Similar in concept to the federal government's Arlington, Virginia station of the 1910s, it was to serve as a local, experimental broadcasting entity. From this highest point in Houston, Peine and Dargan's broadcasts included a daily local time signal from this station starting at 7:00 in the evening. Another broadcast relaying the Arlington station time signal was given every afternoon at 3:00 p.m. Local amateurs of the Houston Radio Club were allowed opportunities to practice transmitting messages by airing broad announcements such as news, weddings and births from the station.[41]

As the 1910s turned into the 1920s, Houston had all the ingredients in place to keep it the state's largest city. Industries such as the oil business now considered Houston its home, and others such as the railroad flourished with the introduction of oil. The creation of the deep port for Houston gave it the defining element to make it a diversified metropolitan area. A post-World War I prosperous economy brought a huge migration of people to Houston, in time to participate and witness the city's first broadcasting boom.

Chapter Six

"Stand by for phonograph concert."[1]

This announcement was printed in a *Houston Chronicle* article accounting the activities of the Houston Radio Club on December 5, 1920. The "phonograph concert" aired from the amateur station, licensed as 5DX, of Ben Still, longtime wireless enthusiast who for years had been the leading supplier of radio equipment in Houston from his Hurlburt-Still Electric Company store. The article went on to proclaim that every Sunday afternoon, "At 3:00 the latest jazz music, marital airs and popular songs would be wafted through the unseen ether" and added, "if you have an acquaintance who has wires strung around the roof of his home don't fail to hear the 'wireless concert.'[2] The Ben Still music broadcasts began with some regularity starting on March 28, 1920 when he made local headlines news and aired an afternoon program followed by a 7:00 p.m. hour-long radio broadcast of phonograph music.[3]

From a wireless apparatus located on Mr. Still's residence at 2520 Beatty Street along the banks of Brays Bayou, these broadcasts originated from station 5DX in the community of Brookline, miles southeast of downtown Houston.[4] Eventually incorporated into Houston, Brookline was formed in 1912, in direct relation to the electric passenger railway system called the Interurban that operated between Houston and Galveston. By the 1930s, the Interurban railway would give way to the automobile industry and its right-of-way is now where Interstate 45 runs south toward Galveston.

Ben Still was probably not as concerned with advancing the progress of radio broadcasting as he was going on the air to endorse his Hurlburt-Still establishment to a scattered audience of amateurs. Most broadcasting

Radio technicians monitoring their broadcast from a Hurlburt-Still station in 1920. *Courtesy of Michael Payne.*

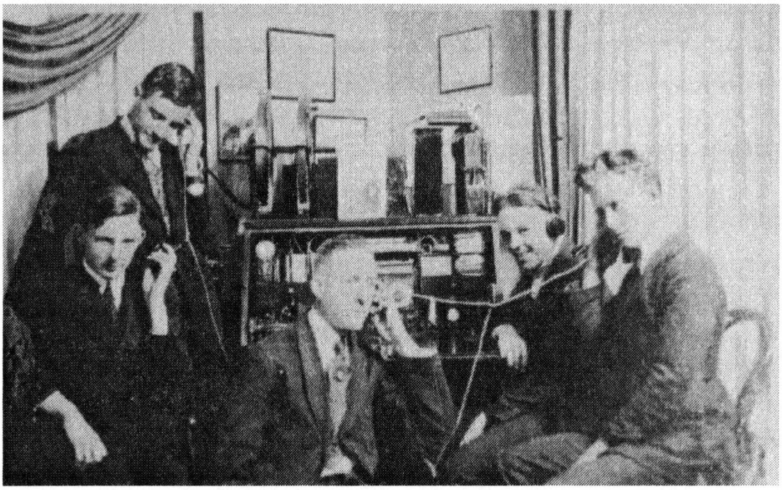

A Hurlburt-Still Company representative (seated, lower left) with Alfred Daniel (center) at his home listening to a broadcast in 1921. Also pictured are Frank M. Austin, (standing, upper left) Ingham S. Roberts, (seated, back right) and Gerald Chinski (seated, front right). In a year, they will become some of Houston's first major broadcasters. *Courtesy of Michael Payne.*

of the early 1920s was conducted as an advertising medium for the broadcaster. In this case, Still was airing radio programs for the sake of his radio department at the Hurlburt-Still Electric Company. The department had now become a large revenue generator for the electrical store that was manufacturing crystal sets and selling radio receivers. As an added attractive to the Hurlburt-Still Company, Ben Still also maintained station 5DL on the store's premises to offer broadcasting demonstrations to curious customers.[5]

Three officers of the Houston Radio Club were employees of the Hurlburt-Still Electric Company during these groundbreaking broadcasts. Acting as retail clerk in the radio section was the club's secretary, Louis Peine. Hallet Worthington, city manager for the club was also employed at the store and was credited in helping design the broadcasting set used by Ben Still. Another member of the Houston Radio Club working at the electrical establishment was John Inglis. A native of Edinburgh, Scotland, Inglis served as a radio instructor in Europe during World War I and later arrived in Houston to work as an automobile mechanic in the Hurlburt-Still service garage. [6]

In Kountze, Texas, a small rural town twenty-seven miles north of Beaumont reportedly received the musical concerts from Ben Still's 5DX.[7]

Things were also on the move for the Houston Radio Club during this time. Just a small group a few years before, the organization had now grown in direct relation to the increasing popularity of radio broadcasting. Instruction in wireless was still being provided and increased in popularity so much that before the end of 1920, a second class was added. Beginners took the easier, low-speed instruction in telegraphy courses while amateurs better versed in telegraphy took the more difficult, higher-speed instruction. The classes were held on Monday and Friday evenings at the YMCA downtown facility.

The year 1920 was also when the Houston Radio Club enacted several measures for its membership. In the spring, the club began publication of its own newsletter called the *Radio Bug*, a take off from the nickname given to amateurs. In December, the club adopted an insignia, a vacuum tube that looked like a light bulb with a flash of lightning in its background. Club members wore this emblem on articles of clothing and printed it on their stationary.[8]

Also in 1920 the club underwent major changes in its membership. Quite uncommon during these early days in radio was a female actively pursuing wireless. But the Houston Radio Club had one in Mrs. E. G. Regan. A native of New York, Mrs. Regan served as a wireless operator during World War I and decided to reside in Texas after the war. Although she lived hundreds of miles away in a small rural community near Fort Worth, Mrs. Regan was considered a member of the Houston club, offering her assistance in their activities.[9] Other changes for the club that year included founding members Alfred Neal Dargan and James Weatherford dropping out of active leadership. Concentrating on his collegiate activities while attending the Rice Institute at this time, Dargan relinquished his duties as club officer and was said to have been influential on the university's broadcasting efforts during the early 1920s.[10] Coming to the foreground for the Houston Radio Club was Alfred Daniel. His years of involvement in Houston wireless and a local World War I hero made him a popular choice to serve as president. Serving as vice-president was Ingham S. Roberts.[11] It was under this leadership that much of these new enactments mentioned occurred for the club.

Houston's radio pioneer James Autry also resigned from active participation in the Houston Radio Club during this time. After graduating from the Rice Institute with a degree in electrical engineering, Autry started his career in the oil industry as an officer in a petroleum company headquartered in Louisiana. Tragically, his life would be cut short. In 1922, at the height of the broadcasting craze, Jimmie Autry suffered from an abrupt case of appendicitis and died shortly after the start of an operation to restore his health. He died just weeks before his marriage to Miss Ruth Streetman.[12]

Group photo of the Houston Radio Club in 1920. Pictured are Alfred Daniel, (front center) Alfred Neal Dargan, (seated, second from left) Hallet Worthington, (seated, third from left) Mrs. E. G. Regan, (seated, third from right) and Louis Peine, (seated, extreme right). *Courtesy of Mrs. Jerry T. Millar*

Years after Autry's death, Harvey Wheeler, a traveling radio technician for the Star Electric and Engineering Company who was later involved in some of the city's larger broadcasting stations up into the 1940s, made a service call to the home of the surviving members in the Autry home during the mid-1920s. Following the repair of the family's radio receiver, Jimmie Autry's mother invited Wheeler to see the equipment Jimmie had used as an amateur. What Wheeler saw was an impressive wireless outfit left the exact way Autry left it-an enduring shrine devoted to the young man's passion.[13]

Eight months had gone by since Clifford Vick was portrayed as a hero for the information he provided to the *Houston Chronicle* about the 1919 hurricane. As prescribed by Vick, the city's largest newspaper agreed to report on-going activities of the Houston Radio Club in return for his exclusive on the hurricane. Articles detailing the club's activities and a full-page photo section were featured in the *Chronicle* shortly following that September day in 1919.[14] But in maintaining an interest in Houston radio during this time, the *Houston Chronicle* also realized the potential of wireless that could be used to the newspaper's advantage.

As president of the Houston Radio Club, Alfred Daniel formed an agreement with the *Chronicle* whereby the radio club would broadcast the newspaper's news releases over the airwaves. At the time, it was quite a unique development. Broadcasting was quickly catching on and the newspaper envisioned a larger audience to be reached. For the radio club, it also provided yet another opportunity for Houston amateurs to learn more about wireless transmitting and receiving. The *Chronicle* had a new advertising medium.

In May of 1920, the agreement came into effect as the Houston Radio Club began transmitting the news as prepared by the *Houston Chronicle*. In exchange, the newspaper agreed to provide a service whereby those receiving the broadcasts and having questions about wireless could send letters addressed to the radio club forwarded by the *Chronicle*. Scheduled at 7:00 every evening, a member from the Houston Radio Club equipped with a powerful sending station would be assigned to transmitting the day's news items.[15]

Known to have participated in this sending assignment were club officers Alfred Daniel and Ingham Roberts. Amateurs as far away as the state of Iowa eagerly waited hearing the regularly scheduled wireless news flashes. Informed ships out in the Gulf of Mexico also waited in anticipation as a Houston amateur reported the current local, international and weather news they would not otherwise have access to. A favorite for many Texas listeners was the state's baseball scores.[16] The shift from the active sending and receiving amateur audience to those only interested in the receiving side of radio transmissions was becoming more apparent as the

Chronicle would feature these wireless reports in the printed press and would remind readers that all they needed to hear the news was a simple receiving crystal set.[17]

The news was aired through telegraphy, still the preferred radio communication used by most amateurs in 1920. While the 7:00 p.m news flashes were being sent, Houston amateurs were asked to abstain from transmitting during this hour so that the transmissions could travel as far as possible.

The *Houston Post* also had a hand in promoting local radio and began indulging in broadcasting at this time. Not to be left behind by its rival the *Chronicle*, the *Post* participated in one of Texas' historic radio feats that has gone down in state legend and lore. Since early 1920, the Rice Institute's radio club had successfully established a wireless communication link with other Texas universities whereby instant sports results could be sent and received immediately following the sports event. By 1921, amateurs at Texas A&M University and the University of Texas conceived the idea of transmitting a play-by-play report of the football game between Texas A&M and the University of Texas played on Thanksgiving Day of that year.

What has been called Texas' first broadcast of a football game, Houston's Louis Peine volunteered to receive instant messages on his radio set at home from Kyle Field in College Station, Texas and would immediately telephone them to the *Houston Post*.[18] Since the play-by-play reports were aired in telegraphy, special codes and abbreviations had to be designated before the game to keep up with the ongoing action. For example, "Texas ball on the Aggies' 45-yard line" was aired as, "TB A 45Y." Another code, "T FP 8Y L," represented "Texas forward pass; 8-yard loss."[19] Along with Louis Peine, Texas stations in Galveston, Greenville, Cladwell, Austin and Waco heard the event that historians have incorrectly reported as occurring as early as 1919 and has been described as the "first ever."[20] Actually, broadcasting of sports events around the nation had occurred before November 23, 1921 by the time local amateurs first aired this Texas football game.

One particular matter discussed at the 1921 American Radio Relay League convention held in Chicago especially caught the interest of Clifford Vick. By now, advances in radio technology reached the point where amateurs could converse with others located overseas through radiotelephony. The American Radio Relay League passed a motion at the 1921 convention to hold a contest for amateurs to break the limitations of contacting Europe. The first amateur in each of the eight districts to make contact with Europe through radiotelephony would be declared the winner in his respective district.

Without question, Clifford Vick had to have been one of the favorites to win in the fifth district. Because the contest was held according to districts, late 1922 was the season designated as the one for the Texas region. This gave time for Clifford Vick and others in Houston months to prepare. Along the way, Vick would achieve other landmarks in Houston radio.[21]

Shortly following the announcement of the contest, Vick received an invitation to head a project few others in the area could achieve: broadcast the proceedings of a city festival. In the years proceeding World War I, an array of citywide festivals came about to replace the popular No-Tsu-Oh carnival that ended in 1916. One of these was the Houston Fair and Exposition, a weeklong festival designed to celebrate and showcase the city resources and vitality.

And what better way to advertise the Houston Fair and Exposition than broadcasting some of its attractions? The Uradia Company, a newcomer to Houston in the business of selling industrial batteries and electrical generators commissioned Clifford Vick to build and operate temporary stations designated to transmit musical concerts from the fair. The Uradia Company needed a unique advertisement ploy, as by this time, the Hurlburt-Still Electric Company was a leader in the city with credit due to Ben Still's broadcasting activities.

Three weeks before the seven-day festival, Vick began airing an open invitation to all mayors and citizens around the nation to come to Houston and join in the celebration. The invite started with a message for "all stations to stand by for important information" (designated in amateur coded lingo as "QRX"). It was then followed by a "general call to all stations" (designated in radio code as "QST"). Vick hoped two stations, in particular, one in Roswell, New Mexico and the other in Little Rock, Arkansas, would copy the invitation and relay it to more outlying stations. Considered some of the most powerful stations in the range of Houston, these two stations would also pass the invitation for stations around the country to listen to wireless concerts airing from the Houston Exposition and Fair in the month of November 1921.[22]

The idea to broadcast the 1921 Houston Fair and Exposition was nothing short of elaborate as receiving stations were installed in various points in the fair's grounds. Houston's leading department store, W. C. Munn, was equipped with one of the broadcasting sets on its fifth floor where a small auditorium was made to house the many visitors expected to witness the broadcast. Another sending and receiving set was placed in the banquet room on the ninth floor of the Stowers Furniture Company. Along with the W. C. Munn's department store, the Stowers Furniture Company was interested in demonstrating a new item they were merchandising called a radio receiver. Located on the fair grounds was a third radio set ready to receive the broadcasts coming from the merchant stores. As the live musical concerts were aired from the W. C. Munn store, the radio set on the fair grounds was converted into a receiving set with a loudspeaker attached to the amplify the entertainment and applauses from audiences at the department stores. Assisting Clifford Vick with this groundbreaking event was Ingham S. Roberts III.[23]

With everything in place, Vick and Roberts aired some of the first live performances to originate from Houston that first week in November 1921. Providing the entertainment were local performers from some of the best musical institutions the city had to offer.[24]

As 1922 commenced, Clifford Vick had the advantages in place to win the competition for being the first person to transmit to Europe. Vick was using his station atop the Carter building, previously owned and constructed by Louis Peine and Alfred Neal Dargan in 1919 for their broadcasting experiments. Vick was a popular radio dealer and builder, and because his station was the highest in Houston, it also served as an invaluable advertisement vehicle for his business.[25]

A series of preliminary tests were conducted to eliminate amateurs in the competition to reach Europe first. With everything in order, Vick passed the screening process and was one of the finalists to represent the fifth district. In late December 1922, the contest began and Houston amateurs hastily sent out their voice messages hoping that they would cross the Atlantic Ocean.

Unexpectedly and quite quickly, a dark horse came from out of nowhere and was declared the first to achieve the feat. On January 5, 1923, Houstonian wireless operator Earl Lester was declared the first in the district and only the second west of the Mississippi River to achieve voice contact across the Atlantic Ocean. An operator in England reported to have received Lester's transmission through wireless telephony.[26] The first to achieve contact in Europe west of Houston was Louis Falconi in Roswell, New Mexico, ironically a close wireless partner of Vick who had helped the Houston operator to relay messages of the 1921 broadcasts from the Houston Fair and Exposition. In all fairness, though, many other elements affected the performance of radio transmissions beside the equipment and operator, especially during this period in 1922 when radio would experience its most profound transition ever with the advent of broadcasting flooding the airwaves. Transmitting continued to be better in the evening hours than the daylight period. Too much radio traffic on the air also affected one's transmission, which is why so often amateurs would sign off when another broadcasting station was on the air trying to reach as many as possible.

Other Houston amateurs were applying the new technique of radio broadcasting. Some merely signed on to air a phonograph record, give weather reports, or simply read articles from the newspaper. But other amateurs took broadcasting into another unique direction. Roy Lay arrived in Houston from Yoakum, Texas to attend the Rice Institute sometime in the early 1920s and was already a seasoned radio operator. He credited his older cousin, Clifford Vick, with influencing him in wireless.[27] Aside from his collegiate studies, Roy Lay utilized his radio apparatus to aid law enforcement agencies.

The City of Dallas implemented a radio station and was issued the first commercial broadcasting license in the state and only the fifth in the nation in September 1921.[28] Designated as broadcasting station WRR, Dallas authorities aired messages relating to fire department dispatches and police emergency calls. In between these transmissions, WRR would air music and a variety of other informations including police reports. One of these daily reports detailed a list of stolen civilian cars reported to the Dallas police. These reports became a statewide favorite for amateurs who would then relay them to other outlying stations and report to the law-enforcement authorities if a stolen car had been spotted.

On at least one occasion, efforts by Roy Lay helped the officials retrieve a reported stolen car.[29] Eventually, Dallas police took the assistance of the amateur seriously as exemplified when they appointed Alfred Daniel "official broadcaster for the state police,"[30] sometime in late 1921 or early 1922.

Acting as a broadcaster for law enforcement agencies was not the only form of broadcasting Daniel was conducting before 1923. Of course, his broadcasting ventures began on a regular basis in 1920 when he was in charge of the *Houston Chronicle* new flashes tapped out in Morse code over the airwaves at 7:00 p.m. As 1921 rolled around, Daniel was pioneering cutting-edge technology and transmitting live voice musical programs on a regular basis from Houston.

Following his service in World War I, Alfred Daniel returned to resume his civilian life in Houston. His years of experience working in some of Houston's largest retail stores inspired him to open his own business as a window trimmer by 1920. But in between the waking hours of designing and installing window displays for retail stores, he pursued his true passion—radio. He spent his evenings presiding over the Houston Radio Club, instructing wireless to new enthusiasts and listening to the more common occurrence of wireless concerts airing from other parts of Texas and the nation.

Neighbors must have looked on with puzzled interest as an elaborate antenna system was installed on Daniel's residence for his ever-expanding radio experimentations. An aerial was mounted on the rooftop of his two-story home at 2504 Bagby Street just west of today's downtown in what is known today as Houston's midtown. Further down his property, another aerial, standing about 60 feet high from the ground completed his antenna apparatus. It was from here in this middle-class neighborhood that practically all of Daniel's broadcasts originated prior to 1922 from his station 5ZX. A lifelong bachelor, who never had any children, Daniel had the free time and area in a spacious home to construct a broadcasting station, complete with a studio.[31]

A few friends and family members gathered at Daniel's home in 1920 or 1921 for what was said to be his first broadcast featuring live performances. Among the party guests were local singers Mrs. Anna Clyde Plunkett, Mrs. Elizabeth Blaffer and pianist Louise Daniel. Secured in a tightly closed room upstairs, the performers sang Brahms' *Saphic Ode* into a microphone that may have resembled more like a modified telephone mouthpiece

Alfred Daniel's station 5ZX equipped with phonograph player, (extreme left) loudspeaker and storage batteries on the floor, circa 1921. *Courtesy of Mrs. Jerry Millar*

that was connected to a radio transmitter. Downstairs, Alfred Daniel sat in front of another radio set attempting to tune in the Brahms' piece as family members and acquaintances eagerly looked on.

Mrs. Anna Clyde Plunkett probably had not finished the musical piece when Alfred Daniel bursted into the makeshift studio shouting; "It works! We heard you just like your were in the same room with us."[32] The broadcast continued with a song called, *One Fine Day* and concluded with a duet from the opera, *Madame Butterfly*.

A visit to the 1921 American Radio Relay League convention in Chicago as a Houston Radio Club representative introduced Alfred Daniel to new theories of broadcasting using unique resources. Upon his return to Houston in September 1921, he decided to apply the theory of transmitting voice over a long distance using only small, low-powered batteries to charge his radio. Few amateurs around the nation had successfully accomplished this feat. Daniel hoped that his experimental transmissions would reach outside Houston with his innovative apparatus. Incorporating a standard transmitter, vacuum tubes, a telephone mouthpiece and

Alfred Daniel at the microphone of station 5ZX and ready to broadcast from his home located at 2504 Bagby St., circa 1921. *Courtesy of Mrs. Jerry T. Millar*

the non-conventional flashlight batteries, he went on the air one evening from his home. He first made the announcement to all that heard his transmission to contact him.[33] Then, he patiently waited.

Two different accounts acknowledge that he was heard. A Western Union telegram operator in Rock Island (assumedly in Texas) heard Daniel's transmission and promptly contacted him. During the response, the Rock Island audience requested the airing of music. Setting up a phonograph player and record, Daniel proceeded to air the song, *Stars and Stripes Forever*. Another account claims a telegram was sent from Victoria, Texas, attesting that the song was heard there as well.[34]

To explain the difference in clarity from a phonograph record and a live human voice heard over the radio, the locally owned James Furniture Company commissioned Daniel to demonstrate a broadcasting of music on January 1, 1922. Among the new home fixtures being offered at the James Furniture Company stores during the early 1920's was a state-of-the art stand up phonograph player called a playerphone and the new centerpiece of the living room or parlor called the radio receiver. Some 25 employees and friends visited the home of J. W. James, president of the furniture store, to hear a specially organized wireless concert.[35]

Another group comprised of Houston Radio Club members gathered at the home of Louis Peine on New Year's Day 1922 to also hear the concert. At 3:00 p.m., a record was cued on the playerphone at the home of Alfred Daniel and was sent over the airwaves to its intended audiences. After the phonograph record was finished, a local singer performed a few selections. The hour-long concert was heard in the nearby towns of Baytown, Galveston and the World War I military base, Camp Logan.[36]

As Ben Still, Alfred Daniel, Clifford Vick and other local amateurs were showcasing entertainment programs for the Houston radio masses; it was quite evident that broadcasting was going to have a profound effect on the American lifestyle. But until the mid-1920s, broadcasting stations were being installed in the most unusual locations. In 1921, a newly constructed automobile service station for the Hurlburt-Still Electric Company served as a studio for the diversifying Ben Still broadcasts. On the corner of McKinney Avenue and San Jacinto Street, the Hurlburt-Still service garage initially served as a studio two nights a week.[37] The advent of featuring live performers made it necessary to have a larger space than what Mr. Still had at his residence in Brookline. Hordes of visitors crowded into the service garage to witness the broadcasts airing on Wednesday and Friday evenings from 8 to 10 p.m. Although the garage had a floor space of fifty by 100 square feet, the studio area was considerably smaller and on many occasions was unable to accommodate the large influx of visitors.[38]

Aside from his radio endeavors from his home, Alfred Daniel also participated in the Hurlburt-Still broadcasts during this period. Daniel would later recall the regular nightly efforts they made to convert a service station into an acoustically acceptable area to transmit music over the airwaves. "Ben Still and I would close the garage doors as soon as the cars left to be repaired were inside, roll out a carpet, set up a mike and start the evening's broadcast."[39] Undoubtedly, the area where the artists performed was divided from the rest of the service garage by a wall of drawn curtains to prevent the music resonating throughout the building and back into the microphone.

Another literary source notes that Alfred Daniel served as announcer to a daily program from the Hurlburt-Still station airing every weekday at 11 a.m. The hour-long airing featured news items, weather reports and a schedule of the evening's musical program. Assigned with a "special land station" license, the Hurlburt-Still Electric Company's station 5DL was allowed certain transmitting privileges not allotted to ordinary amateur licenses. Taking advantage of these privileges, station 5DL chose to broadcast its programs on a frequency of 263 meters (1300 kHz), and a distance on the radio band from the congested 200 meters where amateurs were confined to.[40]

Prior to the war, radio stations were listed in publications by the strength of their transformer. After the war, stations were now measured by the wattage generated by their transmitter. Reportedly, the broadcasts being conducted by the Hurlburt-Still Electric Company were going on the air with a transmitting power of 5 watts in 1921.[41]

A flood of interest in radio broadcasting would reach a fever pitch in the spring of 1922 and inundate the mainstream culture like nothing before. Just a decade before, the future of radio seemed uncertain and even gloomy with the rise of scandalous wireless companies tainting its image and the strict regulation from the federal government to the point that radio was being seriously considered being banned permanently from the private sector. But through the persistence of the amateur, radio in the private sector survived and thrived in the form of broadcasting.

Some eight entities of different persuasions would be granted broadcasting licenses in 1922, a Houston record for one year that stands to this day. More than any other city in the state, these broadcasters would provide an assorted array of programming and would revolutionize how Houstonians would receive their entertainment and news.

Chapter Seven

In April 1922, the Hurlburt-Still Electric Company got the jump on the other Houston broadcasters and was granted the first exclusive broadcasting license in the city. Designated WEV, this license called for the station to air "entertainment and information programs."[1] The station would also be the first and only broadcasting license in the city designated with a three-letter call sign. By the spring of 1922, the Department of Commerce, with so many licenses being issued, would run out of combinations and would start assigning four-letter call signs.[2]

Under the terms of the license, WEV was given two broadcast frequencies. As with all broadcasting stations being licensed as such during this time, WEV was required to sign on at 360 meters (833 kHz) for entertainment programming. But what the Department of Commerce would fail to anticipate was the huge, unexpected number of persons seeking permission to broadcast; causing a frustrating congestion of stations at 360 meters.

But unlike most stations granted a broadcasting license at this time, WEV had permission to also transmit on a second frequency at 485 meters (619 kHz). Stations given this frequency were required to transmit government-sponsored announcements and reports. These broadcasts consisted of weather reports, road conditions, crop reports and, occasionally, stock reports. At 10:00 a.m. every morning, excluding Sundays, a weather bulletin for the Houston area usually flashed from WEV followed by the road conditions of the area. In Houston and especially the surrounding rural areas, a government-generated crop report would follow. For many outlying territories, these agricultural bulletins proved to be essential, as they were often the only means farmers could receive vital information without telephone and telegraph lines. Often, these crop reports would advise when to harvest a certain crop and when to bring it to market to achieve maximum value. So essential were the crop report broadcasts that days or even hours lost in harvesting a crop could have meant the loss of that produce. Radio could provide that immediate assimilation of information previously unavailable to farmers.

A follow-up airing would come on at 5:30 p.m. to update any weather and road condition reports. Most likely, the 5:30 broadcast at 485 meters would also include a schedule and an invitation to all listeners to hear that evening's broadcast of WEV at 360 meters.[3]

Among the radio experts who remained with Ben Still during the transition from amateur license to broadcasting entity in WEV were Louis Peine and Alfred Daniel. An original 1920 participant of the Hurlburt-Still broadcasts, Louis Peine continued to serve as an assistant well into 1922.[4] Meanwhile, Alfred Daniel also began distinguishing himself as a program director for WEV, scheduling and arranging the musical acts that aired from the station.

It was no surprise that the newspaper industry began to take a serious look at broadcasting in 1922. This notion of conveniently entertaining and informing the masses must have been perceived as a serious competitor to newspapers. Historians have noted that prior to World War I, newspapers would often place radio-related articles toward the back pages and treated them with little seriousness. But that would change with the advent of broadcasting. As we have seen, starting as early as 1920, the *Houston Chronicle* took advantage of broadcasting's potential when it agreed to radio its news items in cooperation with the Houston Radio Club. This example served as a precursor to what would occur in 1922.

If one were to scan the Houston newspapers during the year 1922, one would notice quite a different perspective from before. Especially during the spring and summer of 1922, newspapers devoted much attention to radio, often printing more than one article about it in a daily edition. Articles featured everything from the "nuts and bolts" of radio to the ever-expanding broadcasting craze and how it was changing the nation.

By the spring of 1922, "Questions and Answers" columns about radio began appearing in the local papers of the *Houston Chronicle*, the *Houston Post* and the *Houston Press*. Satisfying the public's huge appetite for radio, these newspapers fed that hunger. The *Houston Press* began running a daily section devoted to building and improving reception on homemade crystal sets. Used as a promotional vehicle, the *Houston Chronicle* and *Houston Post* offered manufactured crystal sets for readers purchasing new subscriptions.[5]

From the three major Houston newspapers, the *Houston Post* was probably the last to be considered as launching the city's long-enduring relationship between the press and radio broadcasting. For several years, Roy Garrett Watson had presided over the *Houston Post* with a stern, conservative hand. Watson's father, Julius, presided over the paper before his death in 1897. As specified in Julius Watson's will, Roy Garrett Watson started his rein over the *Houston Post* at the age of 25 in 1917. A refined graduate of Princeton University, the young Watson took the *Post* in a different direction from his father's vision. Oil stock ads were banned. As a devout Christian Scientist, Watson placed bans on patent medicine advertisements as well.[6] Partly for these reasons, the *Houston Post* fell to a distant second behind the *Houston Chronicle* in circulation.

But the paper would make two aggressive moves in May 1922 in an attempt to change its image. On May 8, Watson announced the launching of a second paper that delivered news in the afternoons. Called the *Houston Evening Post*, it provided a slightly lighter view of news than its older sister publication, the *Houston Post*. Not stopping there, on May 18, the *Houston Post* group announced the sponsorship of radio programs from WEV.[7] Under this arrangement, some of WEV's operating functions were given to personnel from the *Post*. Miss

Dewey Harris, music editor for the paper and a gifted vocalist in Houston's music circles, assumed the role of station manager for WEV. Just two years before, Miss Harris, fresh from college and serving as a grade school teacher, accepted a position with the *Houston Post* as a copy editor where she became the first woman in the United States to serve on the universal copy desk of a major metropolitan daily. Acting as WEV's announcer and arranger for musical acts, Miss Harris also answered the station's telephones that were often flooded with calls from listeners with compliments and song requests. As an added feature during her time as WEV announcer, Miss Harris recited a regular evening program for children. Called *Bedtime Stories*, these programs would kick off WEV's evening programming with an assortment of stories and poetry read by Miss Harris from children's literature books.[8]

Undoubtedly, WEV benefited from the sponsorship coming from the *Houston Post*. If any monetary compensation was exchanged, it was reflected in WEV's facility as the station was transformed from a small 5-watt unit to a decent 50-watt outfit at this time.

This marked only the beginning of the *Houston Post's* involvement in Houston radio during these early days. The first weekend that the *Post* began sponsoring WEV's programs, on May 20, management also announced that the paper would begin sponsoring programming from another local station. Alfred Daniel's station 5ZX based from his home began under the sponsorship of the *Post* and started operating under a broadcasting license issued as WCAK. For two weeks prior, Daniel's station aired as WCAK before the *Post* picked it up as its sponsor. Like WEV, Daniel's WCAK ran *Post*-sponsored programs.[9] However, there was a difference in the operations of both stations. Whereas, the *Post* maintained a "hands on" operation at WEV, WCAK operated under Daniel with generally little *Houston Post* staff involvement.

Among the pioneering broadcasts from WCAK at this time occurred on May 28 when Daniel aired what has been credited as Houston's first religious-themed broadcast. Reverend T. J. Windham of the Clemen's Memorial Episcopal Church delivered a fifteen-minute sermon followed by three local singers who sang *Hear Us O Father* and the inspirational, *Whispering Hope*.[10]

Houstonians took to radio with a passion. As WEV and WCAK were signing on, radio-related activity flourished around Houston.

The banners draped over Alfred Daniel's automobile represent the call letters for the radio stations of Ingham S. Roberts (5ZT), Alfred Daniel (5ZX), and Louis Peine (5AE). Also pictured at extreme left is Ben J. Still, owner of broadcasting station WEV and Louis Peine (second from right). *Courtesy of Mrs. Jerry T. Millar*

Telephone calls answered by Miss Dewey Harris poured into WEV, some from as far away as Galveston, complementing the clarity of the broadcasts. Requests for the guest musical artist to perform a particular number were common. Friends would often convene at an acquaintance's home equipped with a receiving set for a radio party. As a member of the Arabia Temple Society, Ben Still placed a radio set at the society's lodge so that his WEV could entertain at social events.[11]

Merchants who previously would not sell electronics were now equipping their stores with radios to attract attention. Stores like the established Levy Brothers Dry Goods and Abe Zindler & Sons installed radio sets on their showroom floors in April of 1922 that were loud enough to attract passerbys on the street. It was common practice that family members and friends to quickly drop off their beloved performer at WEV only to rush to the W.C. Munn's department store a few blocks away and vie for the headphones mounted on the store's radios to hear the concert.[12]

Radio sets were being installed in what, at the time, was the most unusual of places, the automobile. Automobile manufacturers were not equipping cars with radios as of yet, so some resourceful Houstonians took it upon themselves to invent and demonstrate a "mobile radio station."

One such incident involves two historic Houston figures when they were teenage boys. One afternoon, Walter Putnam, Glen McCarthy and Howard Hughes rigged a radio set onto a car and attempted to take it on a mischievous, extended joyride throughout Houston. Along with a receiver, the radio set included a small transmitter and a mouthpiece for airing voice messages. Trained as amateur operators, Walter Putnam and Howard Hughes successfully aired voice messages to all those in radio reach while cruising the streets of Houston. To their delight, the mobile radio station was powerful enough to disrupt transmissions from any station they would pass. In short time, local authorities apprehended the three young men and their vehicle and ceased their activities. Eventually, the three boys grew up and went their own ways. Walter Putnam worked with the Texas Company to conduct research in radio technology for the oil industry. Glen McCarthy and Howard Hughes, two individuals destined to be engrossed in Houston's twentieth century history, became the legendary, colorful oilmen known nationally.[13]

The stormy, rain-filled Saturday afternoon of May 14, 1922, was not the most opportune time to launch what would be Houston's most ambitious broadcasting station before 1925. Nevertheless, on that dreary day, crowds of enthused patrons packed into the Iris movie theater to hear radio concerts. Connected to the large radio receiver was a newly developed Western Electric loud speaker that resonated wireless music throughout the theater and even outside to the street. Because of the inclement weather, the radio concert feature was a partial success on this day. A scheduled concert airing from a station in Detroit, Michigan was unable to be picked up. Instead, the audience had to settle with the local broadcasts that day from WEV and later from Alfred Daniel's station 5ZX that would later be designated as WCAK.[14] Regardless of this minor disappointment, Will Horwitz Jr. must have been excited to know that his theater was equipped to entertain using this new medium called broadcasting.

Houston has little to remember Will Horwitz with. Only those whose lives he touched recall a man generous to his fellow Houstonians during the worst of times, who had quite a character and who was an innovative showman with his chain of movie theaters. A close assistant compared him to P. T. Barnum,[15] ready to captivate the attention of an audience with the finest entertainment product he could give. He surely should be remembered as one of the most colorful Houstonians of the twentieth century.

Born in Benton, Arkansas, in 1886, Will Horwitz Jr. obtained a post-graduate degree from the University of Michigan and entered the newspaper profession. But when he came with his family to Houston around 1917, Horwitz entered a different career as an entertainer and showman. As World War I troops arrived for training at Houston's Camp Logan, he had his first taste as a theater man when he entertained soldiers on the National Guard base with a makeshift theater that featured a regular dose of silent movies and a troupe of vaudeville entertainers. Horwitz knew he had found his calling. In 1919, he purchased a theater centrally located in Houston's downtown entertainment district; a site surrounded by the city auditorium, other movie theaters, restaurants and the area's most luxurious and popular ballrooms. Horwitz renamed his new theater after his daughter, Iris.[16]

Few other Houston entertainment locales in the early days of post- War Houston topped the facilities of the Iris Theatre. Besides a regular schedule of the latest motion pictures, Horwitz's Iris Theatre booked top musical and vaudeville acts for its stage. Already concerned with providing its patrons with a relief from Houston's hot, humid weather, the Iris, for its time, had a modern electrical system that powered an efficient ventilation system to cool its venue. By the early 1920s, its newspaper advertisement often reminded patrons, "It's Coolest at the Iris" at a time when air conditioning was not being utilized on a large scale in public buildings.[17]

Always in search of the latest entertainment crazes, Horwitz saw the potential of broadcasting in the early 1920s. As in the newspaper industry, some leaders in the theater industry perceived this new thing called broadcasting as a threatening opponent. But Horwitz also realized that broadcasting could provide his theater with the added dimension of enticing radio listeners to patronize his entertainment palace. With the thought of "turning your enemy into an ally," Horwitz decided to incorporate broadcasting into his theater.

Houston's P. T. Barnum quickly developed a personal bond with radio. One of his first ventures with radio included the installation of a receiving and sending set on his Dodge automobile. Horwitz had one of Houston's first "mobile radio stations," which was quite a sight in the streets. Sensitive to radio interference, antenna wires were strung from one end of the car to the other and hung over the passengers' compartment. The radio set was attached to the dashboard.[18]

Two years at the Rice Institute and holding down an evening job convinced Francis "Frank" Marion Austin that was all he could endure. He dropped out of the university and once again looked toward the high seas. His love was wireless, ever since he was in grade school, and the exciting adventures it brought him. Since 1920, Austin spent his summer days employed as a wireless operator on maritime ships. His first experience as a maritime wireless operator occurred on a ship plying the Great Lakes, a position held by many young radio amateurs during the still security heightened pose of post World War I.[19]

Now, in 1922, the nineteen-year old Frank Austin longed to sail the seas again after his disappointing try as a college student. But just before he received his orders to report to an Atlantic Ocean oil tanker that year, he accepted a challenging proposition from a good friend and fellow wireless enthusiast. Gerald Randolph Chinski invited Frank Austin's assistance on installing a radio system at Will Horwitz's Iris Theatre. Born at the turn of the century, Gerald Chinski came from Navasota, Texas, where his father, among other things, made a living importing fruit produce. Later, the Chinski family moved to the southwest end of Houston on Francis Street.[20]

Gerald Chinski lived only a block from Louis Peine, and in fact grew up with some of Houston's pioneer radiomen such as Jimmie Autry. However, Chinski's home life during this time was not a happy one. Problems arising from his father's second marriage made the young man yearn to leave home and it was his experience as a wireless operator that gave him the opportunity to do so. Instead of enrolling in college as so many of his friends did after high school, he signed on as a radioman and merchant marine. It was probably during this time as a sailor that he met and started a lifelong friendship with Frank Austin.

By 1921, Gerald Chinski would return to Houston as a well-known wireless operator, brandishing a tattoo on his arm that read, "Sparks"—a sailor's lingo referring to a wireless operator on the high seas.[21]

WEV and WCAK were newly licensed broadcasters in 1922 when Will Horwitz decided to incorporate his theater venture in the growing phenomenon of radio broadcasting.

With an array of antenna wires hanging over the passenger compartment, Will Horwitz (front seat) unveiled his automobile mounted with a radio set in the summer of 1922. In the back seat is Frank Oliver, radio technician. *Courtesy of Houston Post archives, Houston Public Library*

Approached by the city's P. T. Barnum, Gerald Chinski and Frank Austin contracted with Will Horwitz initially to install a large, powerful receiving set for his Iris Theatre.[22] A third wireless expert, Frank Oliver, joined Horwitz's radio venture. Oliver was an older radio operator and an electrician by trade with many scars on his hands from electrical burns he had received during the years.

During the second week of May when the *Houston Post* began sponsoring programming from WEV and WCAK, Horwitz unveiled his theater's radio receiving outfit and from then on initiated a regular schedule of radio-related features from the Iris.

The Iris led the movement of local theaters joining newspapers to be the largest and most influential advocates of radio broadcasting. On May 20, Horwitz staged a radio exhibition at his movie house, giving demonstrations to patrons on how to build radio sets. Along with displays of radio accessories and parts, Iris patrons that day heard a local amateur airing a lecture on radio building and a phonograph wireless concert from the Rice Institute.[23]

Short films discussing radio construction such as "How to Make a Radio Set for $1.00" were also popular during this period at the Iris. Not to be outdone, the Queens Theatre, under the sponsorship of the *Houston Press* newspaper, carried a radio building film, "Make It Yourself."[24] Horwitz was providing Houstonians with what they wanted during this radio frenzy. But it was only an appetizer to the main course he was preparing to serve.

One of the *Houston Post* headlines for June 10, 1922 proclaimed: "WEAY Is Iris-Post Signal."[25] Weeks of preparation from the movie theater and the newspaper had led to this joint venture. For about a month before, Horwitz had enticed moviegoers by bringing radio programs from other broadcasters while the *Houston Post* maintained an interest with WEV and WCAK. Now, Horwitz had the idea of operating a broadcasting station from the movie venue and have it open in time for the Iris' third anniversary in June, 1922. Converting the Iris into a broadcasting station by June was somewhat of an easy task.

For weeks now, the receiving set had been functioning and only some minor details needed to be completed to transform it into a broadcasting unit. The theater's stage and structure provided an excellent acoustically sound radio studio. As evident by the crowds that would gather at WEV and WCAK during their on-air programming, radio broadcasting attracted hordes of curious onlookers. Why not use broadcasting as a revenue generator and charge admission to this attraction? But amid the excitement of being an essential part of a pioneering venture, Frank Austin still longed for the high seas and chose to return to duty as a "Sparks" after the station was completed. Frank Oliver also had other engagements and could not act as a full-time broadcaster and announcer. That left Gerald Chinski to serve as the chief operator for WEAY.[26]

The *Houston Post's* gamble to get involved with radio broadcasting assured the newspaper's place in history. No other Houston newspaper had attempted such a move. In Texas, other print media enterprises rushed into broadcasting. The *Dallas Morning News* was busy trying to be the first newspaper in the state to outright own a broadcasting station while Chinski, Austin and others labored diligently to get the Iris-*Houston Post* station on the air. Perhaps a friendly rivalry ensued between the *Houston Post* and the *Dallas Morning News* as to which of them would enter broadcasting first. Statements have been made that the *Dallas Morning News* was the first newspaper in Texas to own a broadcasting station. This assertion needs to be qualified. In the end, the Dallas newspaper received a license for station WFAA on June 5. A day later, WEAY was licensed to Will Horwitz with some form of cooperation from the *Houston Post.* WFAA would officially begin broadcasting on June 26[27] while WEAY started programming weeks before in early June, which supports the contention that Horwitz's station was the first of its kind in Texas that had some vested interest from a newspaper.

Suspended between two masts was WCAK's antenna system, a design used by many major broadcasting stations of the 1920s. *Courtesy of Mrs. Jerry T. Millar*

Another view of Alfred Daniel's antenna system on his residence. *Courtesy of Mrs. Jerry T. Millar*

Collaboration between the Iris and *Houston Post* produced the finest station in the city. With financial backing that far exceeded what WEV and WCAK had, WEAY was equipped with the most advanced radio components and featured the latest antenna system. Rising above the 115-foot Iris Theatre building was a radio tower more than 60 feet high. Likewise, a similar tower rose from the *Post* building just a few addresses up the block from the Iris on Travis Street. Swinging between the two towers on the buildings were six antenna wires with the ends leading into one wire fastened to each tower. Known as the "T" type, this antenna structure proved to be more effective than the more commonly used single tower design. Radio signals from the "T" antenna were more likely to be distributed to all points, whereas the single tower antenna had an unbalanced tendency to send transmissions in only one direction. This "T" antenna system at WEAY was the same Alfred Daniel used for WCAK, only larger. So as not to create interference with its transmitter, the station's electrical generators were placed on the roof of the Iris. The station's transmitter was located backstage with microphones leading from it to the entertainment stage.[28] Herbert Hoover, then head of the Department of Commerce, signed off on Houston's third broadcasting station during the first week in June. WEAY was on the air by June 9, 1922.

Houston Post authorities approved a WEAY broadcasting locale to be installed on the third floor of the *Post* building. But when the station initially aired on June 9, the small locale constructed to accommodate smaller musical acts and news broadcasts had not yet been completed.[29] Several days would pass before WEAY would begin receiving programming from the studio in the *Post* building.

This joint venture between Will Horwitz and the *Houston Post* to bring ambitious broadcasting to Houston quickly went beyond the boundaries of the studios located in the Iris and the newspaper building. In short time, WEAY's radio technicians devised a communications system to broadcast "remotes" from other nearby establishments. The first site chosen to air a remote from was the city's main auditorium, just a block down from the Iris on Texas Avenue.

Using existing public telephone or telegraph wires, WEAY could air major city events from the auditorium and, if successful, the station's remotes would expand to include other venues. On July 2, just weeks after WEAY signed on, the remote system was successfully tested when Reverend James B. Leavell of the First Baptist Church in Houston delivered a Sunday sermon to his congregation convened at the city auditorium.[30] Wired connections carried Leavell's sermon to the Iris where it was aired to the rest of the city over WEAY.

Other Houston stations would attempt remote airings, too. Later in 1922, WEV engaged in remote broadcasting as it aired a program from the Anton Diehl conservatory, a prominent music school at this time. WEV would also be credited with broadcasting the proceedings of the annual Women's Fair,[31] a convention of attractions for women exhibited at the city auditorium.

When WEAY came on, it achieved what no other Houston station did: it broke through a "great wall of air" generated by Dallas' congested broadcasting activity.[32] Generally, WEV and WCAK, with their lower wattage output, could only penetrate areas in Texas southwest and east from Houston. With so much radio traffic in Dallas, Houston stations had difficulty reaching that city and north of it, until WEAY.

Imagine the jolt Houston listeners received when the much more powerful WEAY came on after WEV and WCAK signed off on the same frequency of 360 meters. With WEAY broadcasting, local listeners were now given a full day of programming. On an average day, broadcasting originating from Houston would begin with WEV airing its government reports of weather forecasts, road conditions and other items at 10:00 a.m., shortly after the Hurlburt-Still stores opened for the day's business.

Radio technician Frank Oliver as he conducts initial transmission testing over WEAY in the summer of 1922. *Courtesy of Houston Post archives in Houston Public Library.*

Throughout the day, the local stations were usually allotted hour-long time slots for broadcasting. WEAY aired most programs during this time, highlighted by the daily noon music concert from the Iris Theatre, a 6:00 p.m. bulletin of news and sports from the *Post* building and an extravagant, nightly 8:00 p.m. musical concert from the Iris stage to cap the day's programs. On days when Alfred Daniel's WCAK aired, the 7:00 p.m. hour continued to be available for his radio concert and news of the Houston Radio Club. Sundays were the least active day for broadcasting, as WEV would not sign on, leaving WEAY and WCAK to alternate presenting religious-themed programs.[33]

An increasingly congested broadcasting scene in Houston did not deter others from rushing to obtain broadcasting licenses. Before the end of 1922, several lesser-known stations would attempt to make their mark in broadcasting. An inventive Houston police officer named George J. Lacy, whose activities in radio started just before WEV and WCAK, was granted a broadcasting license at this time. Born in 1895 in Galveston, Texas, George Lacy came from a simple background. After the hurricane of 1900, the Lacy family moved to Houston where the young George Lacy eventually enlisted in the police department. Known as an ambitious individual, Lacy quickly rose in the ranks and by 1920 was heading the Bertillon section of the department, a predecessor to the finger printing system in which measurements of physical features such as head size and length of arms were used to identify suspects.[34] The Bertillion section was also interested in collecting, maintaining and assimilating criminal data for the public and other agencies.

Officer Lacy's Bertillon section grew interested in the developments that the Dallas police department was licensed broadcasting station WRR and began transmitting reports in the fall of 1921. The statewide success of WRR's broadcasts of stolen cars and wanted fugitives made Lacy realize that Houston's police department needed to keep up with the advancing science of instant communication through radio.

With the broadcasting craze getting ready to take off, Lacy secured the installation of a radio transmitter on the roof of the police headquarters in early spring of 1922. Helping him was local radio technician, George McKenzie Douglas, radio technician for an electrical enterprise called the Tel-Electric Company. As discussed previously in this history, George McKenzie Douglas' interest in wireless started around 1906 in Galveston when he and George Roy Clough designed radio equipment that rivaled those of professional firms such as the American De Forest Wireless Telegraph Company.[35]

Armed with a receiving and sending apparatus on loan from George Douglas, Houston police were now in direct contact with other radio-equipped police departments including Dallas'.[36] Riding on this success and rightfully anticipating the rise of broadcasting, Lacy and Douglas decided to open a business strictly dealing with radio equipment in 1922 at 1118 Capitol Avenue. Proclaiming that their establishment was "the only store in town that exclusively deals with radios" Lacy and Douglas called their store the QRV Radio Company, a reference to the three-letter code used by amateur operators meaning "attention, all stations."[37]

An aggressive salesman, Lacy launched several public gimmicks in conjunction with his QRV Radio Company debut. He gave public lectures and demonstrations at his store on how radio worked by receiving broadcasts from the regular airings of the Hurlburt-Still Electric Company and Alfred Daniel's station. He also donated a radio to the local YWCA on behalf of the QRV Radio Company. During May of 1922, Houstonians in downtown may have seen an unusual sight when Lacy displayed an automobile with a radio set.[38]

Apparently more success was forecasted and the QRV Radio Company opened a second store at 1213 Prairie Avenue, just around the corner from the police station on Caroline at Preston Avenue. Not stopping with the opening of a brand new store and the congestion on the broadcasting frequency at 360 meters, the QRV Radio Company launched Houston's fourth broadcasting station with WGAB during the first week in July at its location on Prairie Aveune.[39] Like WEV, WGAB was housed in the store's facility with a transmitting tower rising 150 feet above the roof. Most likely operating around the same wattage as WCAK and WEV, WGAB nevertheless reached as far away to listeners in Des Moines, Iowa, and Geneva, New York, in early August. Atmospheric conditions were credited for this feat and the QRV Radio Company management anticipated that the upcoming longer autumn and winter evenings would allow for this occurrence to happen more frequently.[40]

Initially, what little airtime was available for WGAB to broadcast at 360 meters was occasionally used for broadcasting two daily weekday musical programs. An afternoon concert at 1:30 p.m. often featured a caller-request format. When opportunity would allow, an 8:00 p.m. concert would end WGAB's broadcast day.

Certainly, the station's most interesting and original programming was its airing of police reports, announced twice a day at 8:00 a.m. and a follow-up transmission of these reports at 4:00 p.m.[41] In the same format as a police blotter, these reports consisted of wanted suspects, stolen cars, bulletins and maybe reports showing how many arrests had been made; data that Officer Lacy would maintain as head of the Bertillion section. Even the police department itself was eager to receive these reports from Lacy's station just around the street corner from it. George Lacy Jr., son of George Lacy Sr., recalled, "My father told me that when the station was ready to go on the air, someone from the police station would poke their head out the window or entrance to indicate to WGAB's operators standing down the block that the police were ready to receive the station's broadcasts."[42] Serving as announcer and operator for WGAB was Ingham S. Roberts III, long time amateur operator, Houston Radio Club official, and experienced broadcaster since 1920.[43]

One of the more curious uses for early broadcasting in Houston came from the local Levy Brothers Dry Goods store. A longtime, well-known dealer of clothing in Houston, the dry goods merchant was issued license WPAN in the fall of 1922. The clothing store was already aware of the potential of radio when it had been equipped with a huge radio receiver earlier that year. By October, the Levy Brothers' WPAN went on the air and was able to situate a daily, morning half-hour time slot. Obviously used as an advertising vehicle, WPAN was devoted solely to different talks geared toward the housewife. Items from WPAN ranged from the "latest women's wear" to the "care of the sick room," a lecture given by a registered nurse.[44]

History would show that WPAN would quickly become an early demised Houston radio station when its six-month license to broadcast was not renewed in early

1923. Perhaps the combination of maintaining a broadcasting station and the congested 360 meters frequency brought an early termination to WPAN. Curiously though, the *Houston Post* continued to print schedules for the station well after the license was not renewed.

Another long forgotten broadcasting license was issued to the *Houston Chronicle* in WFAL. Up to this point, the *Houston Chronicle* had shown little overt interest in getting involved with radio broadcasting, unlike its counterpart, the *Houston Post*. Department of Commerce records show that the newspaper was granted a license for WFAL during the first week in July 1922 but no *Chronicle* headlines appeared nor any other indication that WFAL ever signed on. Only a government and periodical listing survive attesting to WFAL's existence and its subsequent termination in December 1922.[45]

Other entities received broadcasting licenses at this time but were not necessarily interested in maintaining a regular airtime schedule. Veteran wireless operator Clifford Vick opened his radio business called the Vick Radio Construction Company and with it he maintained WSAV from 1922 through 1924. Fred Mahaffey Jr. was a dealer of electrical supplies with his father when he was issued a license for KFCV, a 50-watt station. A student of Houston's Central High School and later a student at the Rice Institute, Mahaffey would occasionally broadcast a short half-hour program starting in the week of February 15, 1923.[46]

Erik Barnouv's *History of American Broadcasting* offers an example of how these smaller broadcasting ventures may have functioned. In 1923, WBBN was licensed to two partners owning an electrical shop that sold radios in Wilmington, North Carolina. When giving a radio demonstration, one partner would go upstairs and turn on WBBN's ten-watt transmitter to air a phonograph record. Meanwhile, downstairs the other shop owner would handle the potential customer as he heard WBBN's broadcast through a store radio.[47] Like North Carolina's WBBN, Clifford Vick's WSAV and Fred Mahaffey, Jr.'s KFCV did not necessarily schedule regular daily programming as they were more concerned with conducting demonstrational broadcasts to sell their radio products.

Chapter Eight

As tenacious as the amateur operators were to developing broadcasting, so too were entertainers anxious to present their forums over the airwaves. In the early 1920s, scores of local musicians, thespians, ministers and luminary hopefuls vied for an opportunity to participate in Houston broadcasting. What they brought to broadcasting set into motion cultural ramifications that still concern us today. Years before America's "golden age of radio" of the 1930s, much of the entertainment genres attributed to this romanticized era were developed at broadcasting's inception in the 1920s.

When WEV came on, conservatory music was the standard of the day. Structured in operatic fashion and "refined and cultured," it often was presented at quaint social gatherings, such as teatime. Known as recital or chamber music, conservatory music was often played by no more than a pianist, singer, and a violinist performing European-styled classical compositions.[1] By 1922, all Houston institutions concerned with instructing music to pupils taught this form of music. Scattered throughout the city's affluent and middle-class neighborhoods, these music schools or conservatories often were held in the instructor's residence.

Since the early 1910s, no other music instructor in town had a greater reputation than Mrs. John Wesley Graham when she became a regular in Houston broadcasting. She came to Houston in 1909 and married John Wesley Graham, president of the Graham Hat Company. An accomplished singer, Mrs. Graham had studied music in Boston, New York, Paris, Milan and Berlin. It was not long after her arrival in town that she became a leader on the Houston music scene. Under her direction, the First Methodist Church, one of the city's most impressive congregations, had a choir that boasted between 200 and 300 voices and was voted the "Best Choir in the South" in the late 1910s. During World War I, Mrs. Graham led traveling choirs that performed for service men at Ellington Field and Camp Logan. Her choirs had traveled throughout the nation and into China, Japan, and South America.[2] Anyone would consider it a privilege to study under her.

Along with her musical talent, Mrs. Graham was flamboyant with luxurious taste and had a large collection of diamonds, which included four stones weighing twenty carats. She exhibited immodest pride in her art when she once graciously commented on a rival voice teacher, "Her pupils sing like hell and she doesn't know it. My pupils sing like hell and I know it."[3]

It was during her tenure as music director of the W. C. Munn's department store's weekly musical concerts that she was introduced to broadcasting. In November 1921, wireless operators Clifford Vick and Ingham Roberts arranged for Mrs. Graham and her choir to perform several wireless concerts for the Houston Exposition and Fair. For the weeklong celebration, Graham's choir provided radio entertainment for patrons visiting the department store.[4] When the broadcasting boom reached a high point in the spring and summer of 1922, her musical protégés were regularly scheduled. At least once a week, local stations, especially WEV, included her performers in their programming. Her pupils performing over radio included Mrs. E. J. Flake, a contralto singer, Miss Corinne Garrison, member of the popular Graham Four quartet, and Mrs. William Uhler.[5] Mrs. Graham's groups were a staple in Houston radio well into the 1930s.

Other local conservatories quickly followed and anxiously began booking times on radio for their performers. One of these aggressive music schools was that of Wilson Fraser and his students. Fraser had just arrived in town in 1919 when radio broadcasting was starting in Houston. He had returned from his World War I service as a machine gunner along the French frontlines. Before his service in the war, Fraser was a classically trained pianist at the prestigious Kidd-Key music conservatory in Texas, and after the war, he remained for a short period in France where he studied and taught piano at a university. A protégé of the renowned piano composer of the twentieth century, Leopold Godowsky, Fraser came to Houston to pursue his career in music. By 1922 he was appointed director of the local Texas College of Music and Art school. During this time, he was a favorite and frequent artist appearing on Houston radio, performing the piano works of Leopold Godowsky and Frederic Francois Chopin.[6] Another frequent performer on Houston radio was operatic tenor singer Henri Therrien who was often accompanied by his wife on the piano and maintained a conservatory at his home.

The Houston Conservatory of Music also showed interest in broadcasting its students. Clarence Hammond headed this conservatory located on Main Street that was a leader in showcasing some of Houston's best new musicians. The Houston Chamber of Commerce appointed Hammond director of the city's conservatory in 1917 and under his leadership, the school began appearing on radio when WEV and WCAK first went on the air in 1922. Houston radio showcased some of the conservatory's finest artists, including an all-male glee club called the Circle H Club that performed short operatic pieces and classical pieces.[7]

The conservatories of Joseph Bragers and Ellison Van Hoose also regularly booked radio performances at this time in Houston. Bragers was a formally trained violinist who began his music school from his apartment in 1922. Under the tutorial of Mrs. Graham, Ellison Van Hoose was a tenor singer who eventually established his own local school in voice training. Van Hoose traveled around the country to perform and for many years was a frequent performer on Houston radio.

Victor Alessandro started a music school in Houston dedicated to instructing young boys in instruments ranging from the brass horns to matching band percussion. To get exposure for his students, Alessandro had them perform over the local stations in big band militia-oriented numbers. Alessandro's son, Victor Jr., would later go to direct the symphony orchestra for the city of San Antonio, Texas.[8]

WCAK's makeshift studios with upright piano and standing microphone at Alfred Daniel's residence on Bagby St. *Courtesy of Mrs. Jerry T. Millar*

Many of the above mentioned musicians and their affiliated conservatories often were aired once a week. For stations WEV and WCAK, much of the responsibility for scheduling musicians seems to have been Alfred Daniel who, of course, also at times served as station announcer and operator. When initially signing on for the day's broadcast, the typical station would begin its day's programming with a series of songs provided by records played on a phonograph.[9] After transmission testing was completed, the live entertainer would perform in the usual time slot of thirty minutes and, in uncommon cases, for an hour. Makeshift studios for these primitive stations were often cramped, insulated for soundproof areas decorated with only a piano, microphone and potted palm trees. Under most circumstances, the local artists were not paid, a practice that continued for decades.

Exposure and notoriety for making a radio appearance seem to have been compensation enough for the performer.

While WEV, WCAK and WGAB tended to air the more cultured music in 1922, WEAY was more daring, groundbreaking and extravagant in its programming. Supported by resources from Will Horwitz and the *Houston Post*, WEAY was dedicated to providing Houston with quality entertainment in a modern, state of the art entertainment facility. Shying away from the higher brow entertainment often aired from Houston's other stations, WEAY tended to schedule radio acts more suited to its audiences who appreciated the silent motion pictures and the vaudeville acts of the Iris Theatre. The station's "conservatory" studio in the *Houston Post* building did air the occasional classical-trained music performer, talks and the daily news reports in the morning and at 6:00 p.m., but the Iris Theatre's main stage provided most of WEAY's programming.

The first summer that WEAY signed on in 1922, Horwitz contracted a six-month stint with a troupe of musicians and stage actors for his theater.

This touring group of performers, led by twenty-five-year-old Tom Stockbridge, became WEAY's first centerpiece act known as the Iris Theatre Entertainers.[10] Since his teenage years, Stockbridge had performed on Houston stages and, unknowingly, was heading a transition into a new medium called radio entertainment. From August 1922 to December of that year, members from the Iris Theatre Entertainers troupe performed over WEAY almost every evening, often finishing the week with a full-blown entertainment production on Saturday evenings. Playing on brass wind instruments with a lively percussion section, the Iris entertainers performed upbeat, ragtime tunes for audiences at the theater and those listening over radio sets.[11] At times, visiting performers joined the Iris entertainers.

Since WEAY's Iris Theatre Entertainers maintained a dominant radio following on Houston's evening broadcasts during these early days, the mornings and afternoons were available for other programming, forums usually geared toward the housewife. While it aired in the last months of 1922, the Levy Brothers Dry Goods Company's station WPAN scheduled its half-hour radio lecture of clothing fashions and women's issues every weekday morning at 10:30. Following this, WEAY's, *Dinner Hints,* a 30-minute talk by Betty Benton Patterson was broadcasted at 11:00 a.m. every Monday through Saturday. Patterson was the editor for the food section called the *Market Basket* and her daily radio appearance repeated suggestions and recipes printed in the *Houston Post* for that day's supper meal.[12]

Immediately following the enticing food talks of Ms. Patterson from the *Post* studios, WEAY aired a noontime concert of musical numbers from the theatre's imposing pipe organ which provided accompanying sound effects and mood music for the silent movies shown at the Iris. Featured during these noontime concerts were the talents of J. W. Watkiss and Hank Hauffe, the theater's pipe organists, who alternated days. Watkiss was the first of these two talents to broadcast an Iris pipe organ concert and was later joined by Hauffe, a recent German immigrant. Pipe organ recitals over Houston radio soon became popular and the organist was often invited to accompany the more elaborate Iris radio concert aired in the evenings.[13] Even beyond the duration of WEAY, the pipe organ concert by a musician employed at a local theater continued to be a popular item in Houston radio.

It is no coincidence that after World War I, America saw a growth and diversification in its music at the same time radio broadcasting was exploding. Broadcasting was exposing more people to different forms of music they could not hear before the creation of the radio concert. Jazz was one of these genres that certainly benefited from broadcasting. Although cultured conservatory music continued to dominant radio programming in the early 1920s, jazz was quickly gaining momentum through radio exposure and was gradually becoming accepted by the mainstream public.

Weldon Jackson Teagarden traveled the Texas circuit as an improvising ragtime trombonist in the early 1920s. More commonly known as Jack Teagarden, the young trombone player was riding the early gestation period of jazz when he joined Peck Kelley's Bad Boys of Houston.[14] Amassing a good following in Houston, Kelley's Bad Boys were experimenting with a new fusion style of ragtime and dance music. This persuaded Teagarden to join the group in the fall of 1921. Their style continued to develop into the style we know as jazz when they were approached to perform at a Houston station in 1922.

On Saturday May 20, 1922, the first weekend that the *Houston Post* began sponsoring radio programs, Peck Kelley's Bad Boys performed a few songs from station WEV. The broadcast proved to be a success and Kelley's group was asked to return at least one more time to WEV later that year. Although Peck Kelley's Bad Boys' lineup was not mentioned in the newspaper for the May 20 radio concert, it would seem quite likely that Teagarden participated in these historic broadcasts for he would remain with the Bad Boys and in Houston until 1923.[15]

More than any other social development, broadcasting helped jazz to become a mainstream accepted style by 1922. Jazz became so popular in Houston at this time that an extravagant benefit called the Musicians Jazz Festival was held in the city auditorium in December. Jack Teagarden and the other Bad Boys performed at this event that jazz historians have credited as "almost certainly the first ever jazz festival on the books."[16] It is also reasonable to believe that this three-hour concert of jazz artists may have been aired over radio as WEAY was already scheduling remote broadcasts of this sort from the auditorium.

In 1923, Teagarden left Houston to play with another jazz group and continued to make broadcasting and music history. In the following year, Teagarden's new group performed in a broadcast that claimed it received more letters and telegrams then any other orchestra had in the entire nation since broadcasting started. The ever-improving and innovative Teagarden would later record with the likes of Louis Armstrong, Benny Goodman, Billie Holiday and would be considered one of the greatest jazz trombonists of the century.[17]

Hundreds of Houstonians attended the city's first radio exposition in September of 1922 held in the city auditorium. From the stage of this entertainment venue many remote broadcasts were transmitted over WEAY. *Courtesy of Houston Post archives, Houston Public Library*

If Houston's Musicians Jazz Festival was aired in late 1922, it certainly left an impression on the city's broadcasters. For within months, many artists who performed at this festival frequently appeared on local radio. A local troupe called the Victory Entertainers performed jazz and ragtime melodies often at WEAY. By 1923, the extravagant Saturday night program from the Iris and WEAY frequently billed the Victory Entertainers as they played with the Iris Entertainers and other touring artists. Houston radio listeners would also at times hear

Jack Sharp's orchestra, which included an exceptional jazz trombone section.[18]

When not operating radio equipment for a broadcast, members of the Houston Radio Club were booked as performers. Several times, Alfred Neal Dargan served as pianist and organist on WEV's programs.[19] Hallet Worthington also played his saxophone over Houston airwaves.

Among those who laid claim to being the first to perform live on Houston radio were Mrs. John Wesley Graham, Louise Daniel, Mary Carson and Charles Painter. Louise Daniel maintained a conservatory and performed on that fateful day from the home of Alfred Daniel when he first transmitted live music.[20] Mary Carson was said to have sung the first entire opera from the Hurlburt-Still broadcasting unit.[21] Coming from the city's Heights suburbs, Charles Painter also claimed to be the first to sing on Houston radio. Painter participated in the broadcasting experiment from the home of Alfred Daniel on New Year's Day 1922 in which a comparison was made on how a live singer sounded differently than a phonograph record over radio.[22]

Few Houstonians took such a liking to broadcasting more than Will Horwitz did. He had a receiving set and mini-broadcasting unit installed in his automobile and, of course, he owned WEAY, Houston's largest and most adventurous station. Weeks after WEAY signed on, Horwitz decided to diversify his resume in radio broadcasting by starting a stint as a radio announcer and talent. Always concerned with workers' union matters, Horwitz first came on the air in July 1922 to announce results of on-going negotiations to resolve workers' strikes being held around the state at the time. His announcements were heard as far away as 115 miles away in Smithville, Texas.[23]

Connections to the city auditorium also made Horwitz one of Houston's earliest sportscasters. On June 12, just days following the initial airing of WEAY, a boxing match between Dave Shade and Tillie "Kid" Herman was broadcasted from the city auditorium. Announcing the welterweight fight was sports editor for the *Houston Post*, Ralph Anderson.[24] After this event, Horwitz must have been totally consumed by the "radio bug" as he decided to man the microphone and announce a series of upcoming boxing matches from WEAY.

Horwitz began pursuing his interest as a radio sports announcer during the rematch between Dave Shade and Tillie "Kid" Herman on July 10. Immediately after every round, an announcer at ringside telephoned the summarized results to WEAY where Horwitz aired the boxing match for all to enjoy.[25] Interestingly enough, the *Houston Post* refused to endorse the bout on the basis that the rematch had been scheduled too closely to the original match on June 12. It was the city's rival newspaper, the *Houston Press,* which sponsored Horwitz's sports-announcing debut. Perhaps it was this reason that allowed the Iris Theatre owner to serve as boxing analyst in a spot usually occupied by a *Houston Post* editor. Nevertheless, this event launched Horwitz's brief career as a radio personality.

Horwitz continued to announce a string of boxing matches through September 1922. His smooth speech and charismatic on-air character soon won him the reputation as the "man with the voice," a distinction proclaimed in the *Houston Post* as it resumed sponsoring the broadcasting of local boxing matches.[26] Quite naturally, in an environment with novices working with cutting edge technology in "uncharted new frontiers," problems occasionally arose. Horwitz was to announce a July 28 fight between Johnnie Dundee and "Kid" Koster, only to be thwarted by burned-out circuits in the transmitter just as the broadcast was starting. Subsequently, WEAY remained off the air that evening until the problems were corrected.[27]

Secular items were not the only programming airing from Houston at this time. Some of the city's more resourceful and aggressive denominations embraced broadcasting as their vehicle to reach the masses.

Typical broadcasting schedule during the fall of 1922. Notice that the stations shared the frequency of 360 meters. *Courtesy of Houston Post archives, Houston Public Library*

FRIDAY'S PROGRAMS

WEAY

Iris Theater Station

(360 meters)

11 a. m.—Dinner hints written by Betty Benton Patterson, market basket editor of The Houston Post.

12 m.—Noonday concert.

2:30 p. m.—Iris theater pipe organ recital.

6 p. m.—Baseball scores and other news of interest.

WEV

Hurlburt-Still Electric Co.

(485 meters)

10 a. m.—U. S. official weather forecast and road condition report.

5:30 p. m.—Final report on weather and roads, together with latest market bulletins from The Evening Post.

---

8 p. m.—(On 360 meters) Enrique Rasoplo, violinist, and Patricio Gutierrez, pianist, will give one of the loveliest concerts yet heard.

WCAK

A. P. Daniel

(360 meters)

7 p. m.—Short musical program and radio club news.

WGAB

QRV Radio Co.

(360 meters)

1:00 p. m.—Noonday concert.

4:00 p. m.—Police reports.

8:30 p. m.—Police reports.

Reverend T. J. Windham's brief fifteen-minute sermon on May 28, 1922, from the Clemen's Memorial Episcopal Church aired over WCAK began a tradition of Sunday broadcasts, mostly devoted to Christian programming. Among the local churches that took an interest in broadcasting, the First Baptist Church was the most often heard during these early days. Reverend James B. Leavell presided over the First Baptist Church with a ministry dedicated to "reaching the world" through international missions and as much public addresses as possible.[28] Naturally, broadcasting fitted well with his church's agenda. Leavell's first involvement in radio occurred on July 2, 1922 when WEAY remote broadcasted his sermon from the city auditorium.

While most Rice Institute students returned home for Christmas break in 1922, Athna Bryan Ellis remained on campus applying his knowledge of radio technology to the construction of a broadcasting station. A native Texan and born in 1897, Ellis contributed his years of wireless experience to manning radio equipment on seafaring vessels in a post-World War I environment. Upon his enrollment into the Rice Institute by the late 1910s, Ellis immediately became a leading figure in directing the radio movement on campus. Not only did he help form and preside over the Rice Radio Club on campus in 1920, but he later became an active member of the more visible Houston Radio Club. When broadcasting was just becoming popular, he participated in a broadcast from campus. Redesigning the university's amateur radio set, he aired a scheduled lecture on "Building Radio Sets" that was heard by interested audiences at Horwitz's Iris Theatre, just weeks before WEAY signed on there.[29]

In late 1922, Ellis had been given the approval from the university to help construct its broadcasting station. Aiding Ellis in this venture were Jimmie S. Waters and H. K. Humphrey, instructors in the campus' electrical engineering department. The Department of Commerce granted a license with the call letters WRAA to the university on November 21. So that the staff and Athna Ellis could devote full attention to the project, it was decided that construction would not begin until Christmas break.[30] Supported by academic resources, WRAA was initially licensed to operate at 200 watts, said to be the second most powerful station in area. Only WEAY bettered its output in Houston. As agreed, WRAA would be a self-contained entity with a transmitter and all supporting equipment located on campus.

In other venues, WRAA may have immediately gone on the air following its construction. However, in Houston's congested radio environment where all local broadcasting stations continued to transmit at 360 meters, the Rice Institute's station could not find an available time slot. It would have to wait and hope that another station would permanently sign off.

The wait would not be for very long as Houston's radio frenzy in 1922 began to subside by the end of the year. In this era when selling airtime to advertisers was considered taboo, stations still had to rely on their own financial resources to sustain themselves. Subsequently, many that entered broadcasting failed to anticipate the high costs associated with maintaining a radio station. Many stations around the nation went under within a year of their inceptions.

The first casualty in Houston's radio scene was WGAB. By late autumn of 1922, it appears that the QRV Radio Company was suffering some major financial setbacks. Internal management disputes and Houston's highly competitive radio market led the city's "only exclusive radio store" to an early demise.

Co-owner George Lacy saw the end coming and started concentrating on other ventures. The Houston police officer who had vigorously marketed his QRV Radio Company and introduced radio communication to the local police department now was diverting his attention to the field of aviation. Lacy purchased two bi-winged airplanes and publicly announced that he would use them for criminal investigation, specifically for the pursuit of wanted suspects. Off the record, though, the entrepreneurial police officer was also considering using his small fleet of airplanes to create a mail courier air service.[31]

By October 1922, the QRV Radio Company was going out of business. That meant that WGAB was also in danger of transmitting its last radio program. The daily radio schedules printed in the *Houston Post* mentioned that "the station was under going renovations,"[32] when in fact, WGAB and the store were preparing to close permanently.

Since its inception, Clarence Hammond realized the potential exposure radio broadcasting would bring to his Houston Conservatory of Music and sought to utilize it as much as possible. Word apparently reached Hammond that the QRV Radio Company was going out of business and WGAB would be abandoned in late October. Since WGAB was licensed in June 1922, the station still had two more months on its six-month permit to broadcast. Seizing the opportunity to reach more mass audiences, Hammond agreed to acquire WGAB and continue its existence under the sponsorship of the Houston Conservatory of Music.

In this period when broadcasting was still in an early experimental stage, individuals and organizations often entered this social phenomenon to explore new ventures. When WGAB was transferred to the Houston Conservatory of Music, other parties acquired a vested interest in WGAB's new broadcasts. There was William H. Whitehead of Whitehead & Bergen, an establishment dealing with marketing mailing equipment and copying documents. Whitehead's interest in WGAB centered on exploring a venture in radiograms, telegrams sent to other parties through radio. Dixie L. Parker, a salesman of the A. T. Vick Electric Company and a dealer of radios was also a participant in what was being described as the "new" WGAB.

Debuting on October 24, 1922, the Houston Conservatory of Music announced that it had acquired WGAB and was prepared to air weekly programs from the school. The station's transmitting equipment was relocated to the conservatory's facility at 2303 Main Street that also served as Hammond's residence. With the transfer of WGAB came its chief operator and technician, Ingham S. Roberts.[33] Still in his early twenties, Roberts had already gained quite a reputation as being an experienced broadcaster in the city.

Depending on scheduling from Houston's other radio entities, the "new" WGAB would air weekly on Saturday or Monday nights at 8:00 p.m. for an hour. Students from the Houston Conservatory of Music were featured exclusively. The hour-long program alternated different students performing works from noted composers and would conclude with a series of radiograms intended to compete with the wire telegram services. But as Hammond may have realized, it was a futile endeavor to maintain a low wattage station when others in Houston were operating more powerful stations and frustrating his efforts.[34]

In February 1923, WGAB terminated all broadcasts, and Hammond's musical groups continued to appear on other Houston stations.

Following WGAB's demise, the Rice Institute's WRAA occupied the slot vacated by the Houston Conservatory of Music station and commenced a weekly program for the 1923 campus' spring semester. After some final modifications, WRAA aired on February 26, 1923 to begin a series of Monday evening programs. Said to be an advertising vehicle for the university, all WRAA programs were operated, managed and performed by Rice students and faculty members. WRAA's hour of programming featured contemporary hits played by Rice students usually on piano, saxophone and stringed instruments. Acting as announcer and operator were

Rice students including Athna Ellis and faculty member Jimmie Waters.[35]

While most of WRAA's programming aired on Monday nights, occasional opportunities arose for campus events to be aired on other days of the week. For example, a Professor Stannes scheduled a Sunday afternoon lecture on literature and poetry in the campus' amphitheater that was aired over the university's station. WRAA had no studios, so the radio operators provided short distance remote broadcasts from different locales around campus, while the transmitter and most of its broadcasting equipment remained stationed in the electrical engineering laboratory. Indeed, the real challenge to air WRAA was dependent on the network of lines connecting microphones stretched across the campus to receive the evening's broadcast. WRAA's first broadcast originated from the university's administration building and later aired a program from the girls' dormitory area. Scattered around the university were radio sets installed for students to gather and listen to the broadcasts.[36]

As quickly as WRAA began broadcasting in February 1923, a decision was made to temporarily cease programming for the upcoming summer break. A grand finale broadcast was aired on May 7, bringing back all past entertainers and ending with the university's alma mater, "For Rice Honor." WRAA promised to return for the fall semester.[37]

Chapter Nine

January 1923 marked an important transitional period for Houston's early gestating broadcasting scene and for station WEAY in particular. While WGAB was transmitting its last program and WRAA was readying to sign on, WEAY raised the standard when it upgraded to a powerful 500-watt station. Taking a temporary leave of absence from his high seas adventures, Francis Austin returned to Houston to aid Gerald Chinski in upgrading WEAY's strength.[1] Among the things Austin witnessed as a maritime wireless operator was a failed attempt by one of his fellow seaman to insurrect a mutiny against the ship's captain. Another occasion, Austin became somewhat of a reluctant hero when he had to desperately send out SOS distress calls aboard a disabled ship out in the Gulf of Mexico.[2]

Many other ambitious stations in larger markets around the nation equipped themselves to transmit at 500 watts, the minimum level required by the Federal Radio Commission to consider it a "high-powered station." With this power, WEAY placed Houston on the international radio map as just days after it upgraded as such a listener in Chile reported hearing a program from the station.[3]

More significantly, the move to make WEAY a 500-watt entity allowed it a rather unique opportunity. During the summer of 1922, the Federal Radio Commission passed a motion allowing for broadcasting stations with a minimum of 500 watts to abandon the already flooded frequency of 360 meters and air at a different frequency of 400 meters (750 kHz).[4] By the start of 1923, some larger stations around the nation were granted this privilege to air at this far less congested frequency. To be able to broadcast from a 400-meter frequency must have been a treasured privilege for broadcasters, certainly for Horwitz, as no other station in Houston had achieved this.

But in order for one to be granted a 400-meter frequency, a station had to meet certain criteria other than a 500-watt station. As specified by the motion, a station had to cease the airing of music recorded onto phonograph records, which was called "canned music."[5] All musical items from a 400-meter station had to come from live acts. In the end, WEAY failed to reach the 400-meter plateau. Though Horwitz's resources had created an impressive WEAY, the ambitious showman failed to reach the condition of no "canned music." For the duration of WEAY's existence, it was occasionally marred by the failure to book a live performer and had to resort to broadcasting an afternoon of phonograph records.

Under this 500-watt power, WEAY continued to air some of Houston's firsts, however. On February 16, 1923, WEAY broadcasted a collegiate basketball game between the Rice Institute Owls and the Texas A&M University Aggies from the city auditorium.[6]

Not detoured by WEAY's eventual inability to acquire a 400-meter frequency, Horwitz continued to diversify his radio enterprise in 1923. Most likely influenced by his former maritime wireless operators, Frank Austin and Gerald Chinski, Horwitz established a commercial wireless station that handled ship to shore messages in September 1923. Prior to this, the only authorized stations permitted to charge a fee for sending and receiving messages were those in Galveston.

Designated station WFO, the offshoot of WEAY was consistently capable of reaching South America, maritime traffic in the Gulf of Mexico and the Caribbean. "I believe a commercial wireless station will be a great benefit to Houston's shipping interests and a good advertisement for the city as well as for my own business,"[7] Horwitz declared concerning his new radio venture. WFO, like WEAY, operated from the Iris Theatre and most likely shared the same powerful transmitting equipment. Here, Horwitz had no competition for no other Houston unit was authorized to handle such a task. Unlike WEAY, WFO was ready for operation twenty-four hours a day. To these ends, fellow Houston Radio Club member Howard Van Eaton, a young wireless operator, was added to man Horwitz's radio enterprises.[8]

After a summer hiatus in 1923, the Rice Institute's WRAA was ready to attempt a project no other Houston station had achieved: broadcast an outdoor sporting event. Heading this groundbreaking feat was returning student Ripley Hanrick who had assisted Athna Ellis during WRAA's initial broadcasts the preceding spring.[9] Faculty member Jimmie Waters also returned to assist Hanrick in airing the Rice Institute Owls' varsity home games for the 1923 football season. Since WRAA's transmitter was stationed in the electrical laboratory and most of the station's programs came from remote broadcasts around the campus, the task of transmitting a football game may not have been such a formidable task as simply as making available an open communication line from the football field. Similar to the layout of the historic broadcast of the 1921 Thanksgiving football game between Texas A&M University and the University of Texas, two radio operators were required to broadcast the football game.

In this case, Ripley Hanrick or Athna Ellis may have been stationed at the football field announcing the game's progress through a telephone line leading to WRAA's "studio" where the second radio operator repeated the announcer's play-by-play details over the air. Among the games aired that season was the traditionally popular confrontation between the Owls and the Texas A&M Aggies on October 20, 1923. During that 1923 season, enthusiastic listeners would call in after the games to tell how they could hear a background of fans cheering and jubilant music from the marching band. So successful were the broadcasts that the freshman football games were also aired that 1923 season.[10]

Though broadcasting local football games went well, Rice Institute administrators had to reach a fateful decision that November. The football season had finished and no more regularly scheduled programming were slated for WRAA. Under the resources given to WRAA, Rice administrators felt that the station had nothing to offer that was not already being done on the local radio scene. The decision was made to terminate WRAA's license on November 20, 1923. A year later, on December 2, 1924, the university renewed WRAA for a brief six-month stint. Said to still be a broadcasting unit, no newspapers mentioned any regular schedule airing from WRAA. Consequently, the station's license expired on May 23, 1925 with little attention.[11]

In a conversation that most likely occurred in early 1924, Ross Shaw Sterling, Sr. posed an enticing question to Alfred Patrick Daniel, "How would we like to have a *real* station?" to which Mr. Daniel responded, "Swell. We'll have some place to play our saxophones."[12] Actually, of course, by this time Daniel already had a venue to air his saxophone on his WCAK. Nevertheless, WCAK's originator was presented with a tempting offer to head up a broadcasting station that would rival WEAY.

Ross Shaw Sterling Sr. was born in 1875 near Anahuac, Texas. After opening a feed store, Sterling entered a banking enterprise and eventually graduated into the oil industry where he became very successful. In 1911, he helped organize the Humble Oil and Refining Company near Houston.[13] Wanting to settle in Houston, Sterling resigned as president of the Humble Oil and Refining Company and in 1924 purchased the *Houston Post* from Roy Garrett Watson.

Immediately, Sterling changed the *Houston Post's* image. Much of the managerial staff was replaced and recent ex-Texas governor William Hobby was named president of the newspaper. The newspaper also merged its morning printing and its evening follow-up, the *Houston Evening Post*, into one daily edition called the *Houston Post-Dispatch*.[14]

But with all the success and responsibility Sterling had, none of it competed for the devotion he gave to his son, Ross Sterling, Jr. As a young boy in Houston, Sterling Jr. became an avid radio enthusiast and regularly attended the weekly popular radio classes taught by Alfred Daniel. Soon, a close friendship developed between pupil and instructor. Amid the classes instructing in radio theory, building radio sets and deciphering wireless codes, Ross Sterling Jr. inspired a dream between his father and his radio instructor to create an impressive broadcasting station, one the city had not seen before.

Ross Sterling Jr. expressed his idea to his father, and the elder Sterling became interested.[15] By 1924, other Texas cities like Dallas and San Antonio had newspapers owning powerful broadcasting stations. A couple of years before, the previous owners of the *Houston Post* partnered with Horwitz to create WEAY. But WEAY's license belonged to Horwitz, and eventually the paper's interest in radio broadcasting faded. Sterling felt that it was time for a Houston newspaper to outright own a broadcasting venture.

Alfred Daniel and Ross Sterling, Sr. reached an agreement that the Houston newspaper needed a broadcasting station to keep up with the times. Despite exorbitant costs, Sterling gave the go-ahead for a *Houston Post-Dispatch* radio station.

Then, tragedy struck. While swimming in Houston's Clear Lake, Ross Sterling, Jr. drowned on April 19, 1924.[16] All thoughts to install the radio station ceased. Knowing that this venture was close to Sterling Jr.'s heart, Alfred Daniel stepped back and decided not to initiate conversation on the project further while a father mourned.

Confronted with the setback of the Ross Sterling tragedy, Alfred Daniel continued to broadcast from his WCAK station. In 1924, he had become a familiar seasoned voice to many Houstonians. His broadcasts still featured well-known performers from the city, news from the Houston Radio Club and the occasional new item such as a reelection campaign speech from a gentleman running for county clerk of Harris County.[17]

A development at this time for broadcasting stations was the use of on-air slogans, a catchy phrase that stated what the station was advertising. Some broadcasting stations were even being granted vanity call letters in conjunction to the station's intentions.

Though no records exist as to when it started, Alfred Daniel used for his WCAK the slogan, "Where eighteen railroads meet the sea,"[18] a reference to Houston, her large railroad industry and her busy seaport. Coined years before, the phrase became a familiar cry of pride for Houstonians by the 1920s. Using this slogan, Daniel's broadcast greetings (or signing off speech) may have sounded like this:

"This is A.P. Daniel broadcasting from station WCAK in Houston, Texas, where the eighteen railroads meet the sea signing on (or signing off)."

Houston's first licensed broadcasting sensation, WEV, also may have had a slogan reminding listeners to visit the Hurlburt-Still Electric Company in downtown Houston for their radio needs. But as ambitious an advertising gimmick was in erecting a broadcasting station, WEV's facilities and programming remained relatively intacted since its inception. In 1924, WEV still aired from a modified service garage decorated with carpeting and potted palm trees. Listeners could still rely on the station transmitting a 10:00 a.m. broadcast of weather reports, traffic reports and other news followed by a 5:30 p.m. update.

About the only thing that changed for WEV and WCAK were their broadcasting frequencies. For most of their existences, they transmitted at 360 meters. But by the summer of 1924, the *Houston Post-Dispatch* began printing radio schedules listing them at 280 meters (1230 kHz), though promotional material from WCAK indicated its frequency at 263 meters (1300 kHz).[19]

It is difficult to give an exact date when WEV and WCAK permanently signed off. Most likely it happened sometime in the second half of 1924. A government listing shows that both stations were deleted from a list of active stations by January of 1925.[20]

Promotional card of WCAK with on-air slogan, "Where 18 Railroads meet the Sea". Note that the station's frequency is listed as 263 meters. *Courtesy of Mrs. Jerry T. Millar*

Perhaps no grand ceremony was held to mark the end of both stations. And maybe the operators of WEV and WCAK had no urgent reason to sign off their stations; just the realization that their broadcasting operations had now become obsolete in a quickly changing broadcasting world. But the fond memories they left behind proved to be these stations' greatest contribution to Houston history. Even decades after their demise, people who listened to WEV and WCAK readily offered a nostalgic memory of how nightly transmissions from a downtown electric store and an impressive bachelor's pad off of Bagby Street pioneered radio broadcasting in Houston.

Frank Austin often returned to Houston when WEAY was being readied for a major change and 1924 was no exception. However, this time, the young Austin was prepared to remain. He had grown weary of his years as a wireless operator on merchant ships and decided to attend night school to pursue an education in electronics which allowed him to assist chief operator Gerald Chinski in what promised to be WEAY's boldest transformation. At first, Frank Austin was placed to monitor Horwitz's revenue-generating station WFO.[21]

Ever the grand showman, Horwitz was posed to unveil two entertainment vehicles in a huge, much celebrated event in 1924. Success followed Horwitz after he opened the Iris Theatre and complemented it with WEAY. Now, in 1924, he wanted to do it again. Already billing it The Texan by June, the second of Horwitz's theaters was scheduled for a grand opening on December 1, 1924. As was reported on June 22 in the *Houston Post-Dispatch*, Horwitz made plans to upgrade WEAY into a 5,000-watt massive broadcasting station. The new station would be associated with his two theaters and would inaugurate its new broadcasting ability when Horwitz's new theater opened during the first week in December. Equipped with a 5,000-watt transmitter, Houston's P.T. Barnum promised that his station would be heard in many parts of the world every night. The June 22 newspaper article also reported that only station WEAF would rival it in strength.[22] WEAF was located in New York

City and was owned by the American Telephone & Telegraph Company (AT&T) that for two years had impressively transmitted radio programs to distant stations through what was called "toll" broadcasting. Through the use of long-distance telephone lines, connected stations subscribing to this service would be assessed a toll fee for receiving WEAF's "feed."

How ironic that Horwitz's proposed station in 1924 would be compared to WEAF, for on at least one occasion, Horwitz connected his WEAY to WEAF's network. The upcoming boxing match for the Heavyweight boxing title in 1924 between Luis Angel Firpo and Harry Wills was getting much attention around the country and Horwitz, the fanatic sports promoter, leased a long distance line from WEAF to air the fight over WEAY. On September 11, 1924, the regular entertainment in the Iris was interrupted as boxing fans jammed the movie theater to hear the match being held at Boyle's Thirty Acres arena in Jersey City, New Jersey. For those unable to get into the Iris, a second location in the city auditorium was connected to WEAF's broadcast through WEAY.[23]

Meanwhile, Chinski and Austin added more off-site venues for WEAY to receive broadcasting programs. Receiving and sending occasional remote broadcasts to and from the city auditorium had proven to be a success and those in charge of WEAY decided that other locations around Houston needed to be accessed for their entertainment programs. Just across the street from the Iris, the Rice Hotel was chosen to receive and air its daily and evening musical events. Since the early twentieth century, the Rice Hotel had been an attractive centerpiece in downtown Houston, featuring all the modern amenities, including a wireless telegraphy set as early as 1908. In 1912, Houston philanthropist Jesse Jones headed a project to renovate the hotel to include a large elegant cafeteria and a beautiful garden rooftop where public evening dances were held.[24]

To connect the Rice Hotel to WEAY, Chinski ran a network of telephone wires into the hotel's cafeteria and rooftop dance hall. He placed microphones at each site and connected them to a relaying system that eventually led out to the public telephone lines and to WEAY. Harvey Wheeler, longtime radio technician for several of Houston's later stations, recalled that Chinski's Rice Hotel wire connections were used by other subsequent broadcasting stations. Remotes from the Rice Hotel became a popular item for Houston stations after 1924 and, as Wheeler recalled, "Chinski's 'pair of wires' into the Rice Hotel would survive and be utilized well into the 1940s."[25]

For reasons undocumented, Horwitz was unable to convert his WEAY into the 5,000-watt juggernaut that he had announced. Stories claimed that perhaps because of a lack of funds Horwitz was unable to receive the radio equipment[26] that had arrived at the local Western Electric dealership to transform his WEAY into a broadcasting behemoth. And what would be crucial to Houston's early broadcasting history, Horwitz also abandoned a new, 500-watt Western Electric transmitter that was to temporarily replace his existing transmitter until the 5,000-watt unit signed on.

In light of this major setback, Chinski's long-distance receiving system expanded to the point that WEAY's programming was dominated by remote broadcasts in 1924. Along with the Rice Hotel; local churches, a university and another movie/vaudeville theater all became linked with WEAY to provide regularly scheduled programming.

By the summer of 1924, as WEV and WCAK gradually curtailed their broadcasts to eventual extinction, WEAY provided most of Houston's radio entertainment. During the weekdays, the 500-watt station began the day's broadcasts with an 11:00 a.m. airing of news and weather reports. These news items were immediately followed by a noontime "remote" of live music from the Rice Hotel cafeteria. Hour-long shows at 2:30 and 6:00 p.m. featured an afternoon of baseball scores and a brief musical recital from the movie theater's organist or any other Houston artist willing to come on. Evening parties and a jovial atmosphere transmitting from the garden rooftop of the Rice Hotel were captured by WEAY starting at 10:00 p.m. and lasting until midnight. From these Rice Hotel dances, the station broadcasted the music provided by the popular jazz orchestras of Lloyd C. Finlay and Max Fink, frequent performers at Houston entertainment venues.[27]

Sunday broadcasts from WEAY provided Houstonians with religious services from at least two of the city's leading congregations. In the beginning of Houston broadcasting, Reverend James B. Leavell of the First Baptist Church appeared at WCAK and WEAY to give short sermons. But by 1924, remote broadcasting brought the station to him and allowed his church, located on Lamar Street, to air entire Sunday services at 11:00 a.m. sometimes followed by a second, hour or two-hour service at 7:00 or 8:00 in the evening.[28] On alternating Sundays, WEAY aired services from the First Methodist Church by Angie Frank Smith, who came to Houston in 1922 to head this church on Main Street at Clay[29].

However, not all Sunday programming was of a religious theme. For a few months in 1924, a series of programs were received from the Prince Theatre in downtown. Although a direct competitor to the Iris, WEAY, nevertheless, picked up late night Sunday vaudeville comedy acts from this entertainment facility. After the Prince's Sunday afternoon showings of silent movies, the 9:30 p.m. program of live musicals acts and comedies was transmitted over Houston airwaves until its completion at about 10:45.[30]

Special events endorsed by Horwitz would at times preempt the regularly scheduled programming. For example, during the week of October 31, 1924, WEAY temporarily ceased broadcasting from the Rice Hotel and aired programming from other various downtown locations. When Horwitz opened his dining and entertainment club called the Milwaukee on November 3,

WEAY broadcasted the entertainment and proceedings. The following evening, WEAY broadcasted a citywide party of music and dancing in honor of the United States presidential election night from the city auditorium. On the next night, a minstrel show held at the city auditorium at 8:15 p.m. was aired.[31]

Logistically, Chinski and Austin's most difficult remote broadcast may have been the popular collegiate football game scheduled on November 1, 1924 from the Rice Institute. Never before had WEAY technicians attempted a remote broadcast from outside of downtown. But this game between the Rice Institute Owls and the Longhorns of the University of Texas was getting much attention and promised to amass a huge radio audience around the state. Even though the Rice Institute still had WRAA licensed but was not using it, a decision was made that WEAY could provide better reception for the statewide fans listening to the game. Those prepared to listen to the "play by play" included a *Houston Press* sponsored party in downtown.[32]

Traversing several miles of communication lines between the football field and WEAY, the game was aired on the afternoon of Saturday, November 1. WEAY's operators received assistance from Rice student Ripley Hanrick who served as the on-air announcer and had served in the same capacity the previous year when WRAA aired football games. Days later, fans from Dallas and other areas sent notice that they heard the game.[33]

One overriding agenda for Horwitz's broadcasting efforts was "distance"- how far his WEAY could reach. Unable to transmit at more than 500 watts, WEAY's technicians directed their attentions in late 1924 and early 1925 to achieve a consistent broadcast radius that extended beyond the borders of the United States. Before, WEAY's transmissions concentrated toward one direction and were often heard in South America, but seldom very far north of Texas due to so much broadcasting traffic in the nation. Then in early 1925, improvements to WEAY and the right atmospheric conditions created a situation that had eluded the

Houston station before. In March 1925, WEAY's signal achieved a 4000-mile broadcast radius that was received as far north as Alberta, Canada, and south to Guatemala. Calls also came in from South Carolina and a ship in the Pacific Ocean to report that the broadcast was heard. What they listened to that night from Houston was a Sunday evening sermon by Claude L. Jones of the First Christian Church, another local congregation that was added to provide "remotes" to WEAY at this time.[34]

As with WEV and WCAK, records are vague as to when WEAY performed its last broadcast. According to Frank Austin's license as a radio operator, he was assigned to WEAY until June 1, 1925,[35] which is perhaps the only surviving record that gives a definitive date to the station's last day. WEAY left behind quite a history, originating from a Houstonian with a vision who became a patron hero for some and an abrasive figure for others. The station benefited from resources the likes that no other Houston station before 1925 had. It remained Houston's most powerful and ambitious radio outfit in its three years of existence. It was a link from the infant broadcasting boom of 1922 to where broadcasting was heading, a full-blown medium.

Chapter Ten

For determined men, dreams are often difficult to abandon, even when they have been extinguished by personal tragic events. During the first months of 1925, Alfred Daniel saw an opportunity to revive the vision that had been marred by the untimely death of Ross Sterling, Jr. By March of 1925, Houston had only one station in WEAY that was regularly scheduling broadcasts. The year before, Alfred Daniel had abandoned his WCAK and felt it was time for the city to have another powerful broadcasting station. In his urges to Ross Sterling, Sr. to consider building a station, Daniel may have mentioned that the *Houston Press*, a competitor to the *Houston Post-Dispatch*, was sponsoring programs over WEAY[1] and that it was time for Sterling's newspaper to compete through its own station. Also, an important Houston event was approaching which would be the perfect opportunity to unveil the new station.

Houston was selected to host the twentieth-first gathering of the Associated Advertising Clubs of the World in 1925.[2] The year before, Houston leaders worked feverishly to convince the international group, dedicated to discussing international advertising and commerce matters, to convene here. They succeeded, and starting in the first week of May delegates from the Association arrived in Houston where for several days the group would hold most of its activities in the city auditorium. Among the dignitaries attending the gala was the governor of Texas and Herbert Hoover, then head of the United States Department of Commerce that issued licenses for broadcasting stations in 1925.[3]

With Houston receiving international attention, Daniel recognized this to be an excellent opportunity to debut a broadcasting station that would transmit the proceedings of the Association to the world. Already a seasoned broadcaster, Daniel knew he had the ability to pull it off. However, his biggest hurdle was to approach Ross Sterling, Sr., a man still grieving, to finance the project. It is uncertain how Daniel approached Sterling with his idea, but suffice to say, the *Houston Post-Dispatch* owner gave his approval—with just three weeks before the convening of the Association![4]

Just three weeks to install a high-powered broadcasting station left Daniel little margin for error. Fortunately, he knew where to go.

Word reached broadcasting expert G. E. Zimmerman about ninety miles east in Beaumont that the *Houston Post-Dispatch* was hurriedly planning to install a broadcasting station and his assistance was crucially needed. Nicknamed "Shut'em down Eddie," Zimmerman was the engineer who would "cut the switch" to end the day's broadcast at some of the Texas stations he helped install. Zimmerman came from a family in the Dallas area that owned an electric store and began selling radio sets to amateurs and broadcasting buffs in the early 1920s. He had a hand in building Dallas' historic station, WRR, and later was a major player in installing WFAA for the *Dallas Morning News* in 1922. His reputation grew as he left WFAA to help design and operate Beaumont's KFDM that signed on in October 1, 1924.[5]

"Shut'em down Eddie" accepted the challenge and came to Houston to help sign on the *Houston Post-Dispatch* station. Along with Zimmerman came Charles F. Baker, one-time chief operator of WFAA,[6] to accept a position in the new station. Now that a core of some of the state's most qualified radio technicians had been assembled, the next major concern was getting together the equipment and having it installed within twenty-one days.

Accounts vary as to how the *Houston Post-Dispatch* got the necessary equipment in time to broadcast the proceedings of the Associated Advertising Clubs of the World. A common notion is that the station came out of the dissolution of WEAY.[7] While there was some truth to that, the fact remained that WEAY stayed on the air for a couple of months after the new *Houston Post-Dispatch* station would sign on. A more accurate account from a reliable source provides a slightly different tale of how Ross Sterling's station began airing.

When Horwitz was planning to launch a 5,000-watt station in 1924, he had also purchased a 500-watt transmitter to replace the one being used at WEAY. As it turned out, Horwitz was unable to realize his broadcasting mammoth, and he also abandoned the 500-watt transmitter that would upgrade his existing station. It was also at this time that Ross Sterling, Sr. and Alfred Daniel were in the planning stages to start a station when Sterling lost his son. As recounted by Horwitz and Zimmerman, Sterling acquired the 500-watt transmitter for the *Houston Post-Dispatch* station that Horwitz deserted a year before. [8]

In 1925, technicians arrived at the warehousing location to find the Western Electric 500-watt transmitter still intact and ready for operation. The transmitter was transferred to the *Houston Post-Dispatch* publishing plant located east of downtown at the corner of Polk and Dowling Streets, where the newspaper was actually printed. With little time remaining for their May 9 deadline, no new building could be constructed for the station. A large hole was cut through the roof of the publishing plant to accommodate a stairway and the broadcasting equipment. From here, the studio and radio mechanical rooms were installed on the publishing plant's roof.[9]

Like WEAY, the *Houston Post-Dispatch* station used the "T" antenna system, a design used by the major broadcasting outfits at the time. The tallest antenna tower rose to 140 feet above the ground. The second tower was attached to the plant's roof and rose 100 feet above the building. Suspended and swinging between the two towers was the antenna, 246 feet of sensitized wire. Based on stations equipped similarly, it was anticipated that under favorable atmospheric conditions, the station would have a broadcast radius of 9,000 miles. The station was complete and ready to broadcast at an estimated cost of $25,000 in 1925.[10]

In selecting the station's call letters, Alfred Daniel wanted the identification to reflect Houston, like his previous WCAK's slogan. After deliberating what the city's major resources were and reviewing available four-digit call letter combinations, Daniel was granted KPRC, which stood for "Kotton, Port, Rail Center." The station's management also agreed that Daniel would be able to sign on the air with, "Station KPRC airing from Houston, Texas" and continue using the slogan he used at WCAK with "Where eighteen railroads meet the sea."[11]

KPRC's first studios on the roof of the Houston Post-Dispatch publishing plant in 1925. *Courtesy of Houston Post-Dispatch archives, Houston Public Library*

For Alfred Daniel, it was a dream come true. He now was able to pursue his passion on a full-time basis. For Ross Sterling, Sr., KPRC served as a memorial for the son he lost the year before and who had inspired the idea for the station.[12]

"Hello folks, everywhere" was credited by David McComb's *Houston: The Magnolia City* as the first official words spoken, most likely by Alfred Daniel, over KPRC on the evening of Saturday, May 9, 1925. The first guest speaker was Reverend William D. Ryan of the South End Christian Church who invoked a blessing and prayer. Other speakers who came on the air that evening to offer congratulatory remarks were Houston's mayor Oscar F. Holcombe and former Texas governor William P. Hobby, now serving as president for the *Houston Post-Dispatch*. Also on the program that night was Lou Holland, president of the Associated Advertising Clubs of the World, to introduce some the musical entertainment from the convention that would be aired.[13]

The open-air, platformed stage built on the roof of the *Post's* publishing plant to accommodate musical groups too large for KPRC's studio came in handy that night. The twenty-five-piece Royal Scotch Highlanders from St. Petersburg, Florida performed on the KPRC stage under a starry night. Arranged by Ross Sterling, Sr., the forty-five piece Humble Oil Company band came on next to perform. The evening program was closed by the seventy-five piece First Garrison band of Mexico sent as a goodwill token for the visiting convention by the president of Mexico.[14]

The next day, KPRC received its programs entirely from remotes, which made a hectic second day of broadcasting for head of the station's remote control devices, Grimes R. Walker. Starting at 10:45 Sunday morning, broadcasts came from the First Christian Church that included religious services and a speech from a former president of the Associated Advertising Clubs. Sunday afternoon programming came from concerts and speeches to entertain the proceedings of the visiting convention meeting at the city auditorium. Among those providing music from the city auditorium and heard over KPRC was a choir composed of 100 African-American young men and women from Prairie View University, possibly the first black performers to be heard over a Houston station. Pastor of the First Christian Church and a recent Sunday regular on-air preacher for station WEAY, Claude L. Jones, closed KPRC's second broadcast day with an evening service.[15]

Ray L. Dudley, vice-president and general manager of the *Houston Post-Dispatch,* had the monumental task of getting everything together and putting the station on in a short period of time. Of course, he succeeded, but he had his doubts initially. "I thought KPRC was too expensive a luxury,"[16] Dudley coordinated the purchasing of the equipment. During the planning stages, he concluded that KPRC would have an operating cost of $30,000 to $35,000 a year, an extreme cost in 1925.

Ray Dudley searched for ways to somehow offset the projected costs to maintain KPRC since, for the most part, selling airtime to advertisers was still unacceptable etiquette for radio broadcasters. Going on the lead of the old *Houston Post* three years before, Dudley decided to run a campaign of issuing a free, inexpensive crystal radio set for every new subscription sold. They would be "killing two birds with one stone," as the *Post-Dispatch* would get a new reader as well as promoting a new listener to KPRC.

Dudley authorized the purchase of massive quantities of small crystal sets at wholesale prices as KPRC was being built. When KPRC went on the air, the free crystal set campaign also went into effect. The newspaper carried large ads stating that anyone in the family could listen to KPRC with this crystal set. More than 12,000 sets were said to have been issued in this campaign, a figure that made local radio dealers nervous. It was not until after a meeting between *Post-Dispatch* officials and radio dealers that the issue was settled assuring that the campaign would only be temporary.[17]

Assigned a frequency at 296.9 meters (approximately 1010 kHz), KPRC initially commenced scheduled programming with daily broadcasts at 12:00 noon with an airing of weather and road reports. The station would sign off and resume programs at 6:00 p.m. with the final show ending at 10:00 p.m. Then in July of 1925, the station expanded its broadcast day.

Standing at the microphone of KPRC are Alfred Daniel (right) and the station's first general manager, G.E. Zimmerman, in 1925.
Courtesy of Houston Post-Dispatch archives, Houston Public Library

At 10:55 a.m., the station would go on the air to carry time signals transmitted from the United States Naval Observatory in Washington, D.C. The receiving of time signals enabled KPRC technicians to test the transmitter before local programming was aired. For five minutes, Houstonians could hear a "countdown," as every minute would be announced. With the stroke of 11:00, the time signals would be interrupted with an extended "buzz" sound followed by the local reports on weather and road conditions. Afternoon programming would then commence and lead to an evening of broadcasts. As a friendly policy immediately adopted and honored until 1928, the station designated Fridays as "silent night" and did not air on that evening as an opportunity for Houston listeners to receive other long-distance stations without interference.[18]

Entertainment over KPRC offered Houstonians some returning familiar voices as well as talents never heard before on radio. Events that had been aired before by previous Houston stations were resumed by KPRC and often with better production. With virtually no other Houston broadcasting station in existence in 1925, the station had unlimited access to air any Houston event.

Program director Alfred Daniel scheduled many musical acquaintances he had met over the years to perform at KPRC. Anne Clyde Plunkett, choir director of the St. Paul's Methodist Church and one of the first performers to sing on Daniel's station at his home, sang with her choir at KPRC on May 27. Another performer for the *Post-Dispatch* station was Mary Carson, the operatic soprano singer credited as being the first to perform an entire opera over radio from station WEV. European-trained classical pianist Wilson Fraser returned to radio performances and performed often at KPRC. The area's most respected conservatories such as the Houston Conservatory of Music also continued their radio gigs.[19]

Militia-style marching bands from the local fraternal order of the Arabia Temple Shrine and the Southern Pacific Railroad marching band became frequent radio performers. At this time, jazz groups such as Eddie's Syncopators and Blanchard's Dance Orchestra made radio appearances to the delight of wireless fans. Early KPRC appearances were also made by students from the Rice Institute performing some contemporary tunes just before the students were dismissed for their summer break in May.[20]

One of the major breakthroughs for KPRC in 1925 was the inclusion of black artists on its schedule. Beginning with the Prairie View University choir on May 10, a series of African-American artists were booked as much as twice a week for a period in 1925. Black musicians such as violinist Russell McDavid, the Pitt's Ninth Regiment Brass Band and the Sims Joyland Orchestra of Houston performed half-hour sessions. On July 22, the Olympia Male Quartet, Davis' Louisiana Jazz Band and the Coleridge-Taylor Choral Club provided an evening of black jazz and Dixie tunes.[21]

Other artists got "discovered" on KPRC and became radio staples in Houston for many years. Dr. Earl Shay was a chiropractor by day and a tenor singer by night on radio. Well-received by radio audiences, Dr. Shay sang weekly on KPRC and later stations for years. From KPRC's inception, Frank Tilton was a weekly regular. Born blind, Tilton was a gifted pianist and was soon nicknamed the "wonder boy." A radio following quickly developed for the young man and often the station would focus an evening's programming around him to showcase his talent.[22]

For many weeknights, usually at 5:30 or 7:00 in 1925, Judd Mortimer Lewis provided an hour of fun for children. Born in New York in 1867, Lewis was in the newspaper industry when he arrived in Houston in 1893. He joined the *Houston Post* in 1900 but left when Roy Garrett Watson assumed ownership of the newspaper. He later rejoined the *Houston Post-Dispatch* when Ross Sterling purchased it, and his humorous prose stories entitled, *Patsy Kildare* and *Jubilee's Pardner* became daily features. He soon developed a following with children. When KPRC signed on May 9, Lewis was one of the first on-air speakers that evening and was offered his own hour-long program. Known as *Uncle Judd's Kiddies Hour,* his nightly program featured readings of children's poetry, literature and entertainment. "Uncle Judd" became an endearing radio figure for a generation of Houston kids and would eventually go on to write several books of verse.[23]

Patterned after WEAY in its last months of existence, KPRC's Sunday programming centered on morning and evening religious services. Returning on the air was the First Baptist and the First Christian churches and would alternative programming. Soon, other local denominations were allotted airtime. Not to be outdone by other churches broadcasting services, Reverend Claude L. Jones of the First Christian Church added a session of Bible study classes to his programming.[24]

On occasion, KPRC would air out-of-the ordinary items. On May 19, a psychic named Eugene Davis was invited to make on-air predictions and answered questions from callers, which flooded the station's phone switchboard. Questions posed to the radio psychic ranged from money issues and love interests to callers asking advice in locating missing or estranged family members. Unable to fill all the requests, Davis returned only days later for another session of questions.[25]

Improvements in communications technology allowed for KPRC to broadcast sporting events previously never aired in Houston. The city was represented in the Texas League by the Houston Buffaloes, a minor league baseball team that traveled the Texas and Louisiana circuit to play other cities.[26] With the Buffaloes' 1925 season already in progress, KPRC aired the team's away games starting in July. This meant sending out a station engineer with the team and securing an available long-distance telephone line at the game's venue.

Equipped with the proper resources, a KPRC technician left Houston to catch the Buffaloes as they prepared to play the Panthers in Fort Worth, Texas, on a Sunday afternoon, July 19. After a telephone line from the Fort Worth baseball field was established, the play-by-play was relayed long distance to the station in Houston where a second announcer recounted the results over the air. Between innings, a second open microphone was located in the station's studio where pianist Frank Tilton performed requested songs for radio fans. A call came in for Tilton from Houston's Ellington Field offering the radio pianist an opportunity to ride in an airplane. Cautious of flying, Tilton responded that he would "play a number for the listener on the piano but did not hanker to learn to play the harp just yet."[27]

For several days after July 19, KPRC followed the Buffaloes on their road trip and interrupted regular broadcasts to air the games against the teams of Shreveport, Louisiana and Dallas. The Buffaloes would end the season with a record of eighty-seven wins and fifty-six losses.[28]

More comfortable with its facilities, the station conducted more "feeds" and "links" from long-distance events. The Fort Worth Panthers won the 1925 Texas League pennant and had an opportunity to play an Atlanta, Georgia, team in the much-celebrated Dixie Series championship in early October. KPRC was there to bring the action to southeast Texas. Afterwards, the major leagues' World Series was aired for Houstonians to hear. The Associated Press allowed for broadcasting stations around the nation to link to its wire service as it carried the baseball games. Sports writer for the *Houston Post-Dispatch*, Bruce Layer, received the games over the Associated Press wire and recounted the plays over KPRC's microphone. Having announced the Dixie Series the week before, Layer provided game results and player analysis for the series as the Pittsburgh Pirates defeated the Washington Senators in seven games.[29]

Sporting-events announcing in 1925 were a far cry from 1921 when the first sports were transmitted over radio through dots and dashes. For the Dixie Series game on October 8, Layer provided a pregame analysis on the players and game. While play was in progress, Layer kept a box score of the batter to announce as the player came up to the plate.

Phone calls came into KPRC's P.B.X. switchboard operator to compliment the series' coverage. From the Martin Lumber Company in Port Arthur, Texas came comments like, "Reception fine. Lumberyard turned into receiving station. All business suspended and all prominent citizens of Port Arthur present." A Mr. Mills of Hempstead, Texas called in to say, "Five hundred fans listening and thanks for service."[30]

A far less logistical task was airing all home football games of the Rice Institute Owls in 1925. The station simply installed microphone and established telephone connections in the stadium's press box to the enjoyment of local, collegiate football fans.[31]

KPRC's inception year of 1925 proved to be an exceptional start for the station.

Chapter Eleven

"Keep Fooling Vid' It," a self-appointed slogan given by the operators that derived it from the call letters appointed to a small broadcasting outfit they maintained called KFVI. When this station signed on in early 1925, the first "performers" were actually horses that were corralled in stables that also served as a broadcasting station. An amateur operator from Long Beach, California sent word that he had heard these four-legged "performers" when Jacob "Dad" Schwarz of KFVI hung a live microphone in the stables.[1]

In 1921, administrators of the Texas National Guard reorganized units of the state militia to be in better readiness to serve Texas and the nation in times of emergency. From this reorganization effort was created the 56th Cavalry Brigade, a unit headquartered at a horse stable facility just west of downtown Houston at 305 Sabine Street along the banks of Buffalo Bayou. Recruits for this unit were trained in everything from firearms and stenography to horseback riding and blacksmithing. Parading in some events around Houston and dressed in military regalia, the 56th Cavalry Brigade quickly became a popular unit.[2]

The only Houston station scheduling regular programming was WEAY when a license was granted to the 56th Cavalry Brigade to maintain a broadcasting station a couple of months prior to KPRC signing on. Initially, all of the station's equipment was housed in the brigade's livery stables.

What interest would a militia unit have for erecting a broadcasting station devoted to airing entertainment programming? Surviving records do not provide a clear answer. Radio for the masses in 1925 was perceived as a relatively still untapped realm with many undiscovered possibilities, and the 56th Cavalry Brigade had an interest in radio communication. By 1924, the brigade was equipped with a powerful, portable transmitting and receiving wireless set for field exercises. Recruits were encouraged to study radio communication.[3]

KFVI was licensed as such in March of 1925 and with a maximum output of only 10 watts.[4] Under this wattage and a congestion of other broadcasting stations, KFVI would be pressed to reach Galveston under normal circumstances. Reaching points outside of Houston did not seem to be a concern of the 56th Cavalry Brigade, though. Instead, KFVI concentrated on broadcasting to Houstonians and those citizens who served in the Texas National Guard, offering them an introduction to radio theory in the convenience of their homes. Broadcasting still provided an excellent attention getter for any organization and KFVI directed invaluable attention to the brigade.

Of all of Houston's early stations, KFVI, in its first year, received the least attention from the local press. The broadcasting euphoria of 1922 had dwindled by 1925 as fewer groups were inclined to operate a broadcasting station. Not only that, but KFVI was a low-powered station with no sponsorship from a newspaper in 1925. Subsequently, no printed schedules survive to attest if the station aired on a regular basis in its first year. One can guess though that the station aired programs infrequently and still served more in the capacity as an amateur-broadcasting outfit in 1925.

On February 17, 1926, the *Houston Chronicle* announced, "Station KFVI To Take Air on Thursday," while the *Houston Post-Dispatch* printed the next day, "Cavalry Brigade Radio Station To Go On Air Today." Both newspapers were either given erroneous information or were running a slightly misleading story. The fact was, of course, that KFVI had been licensed almost a year before. The February 1926 articles also stated that the station would "broadcast daily from 10:00 a.m. to noon and from 2:00 p.m. to 5:00 p.m." The newspaper articles mentioned that programs from KFVI consisted of "lectures of travel, economic and civic subjects."[5] By the summer, though, the station expanded its programming to evenings and, on occasion, before 10:00 a.m. at the frequency of 239.9 meters (approximately 1250 kHz). Although the station transmitted from a different frequency than KPRC, its broadcasts were scheduled carefully so it would not share airtime and cause interference.[6]

The *Houston Chronicle* showed little overt interest in participating with broadcasting, except for the licensing of WFAL, which apparently was aborted before it came on the air in 1922, and the license for a portable transmitter in October of 1925, designated KFYJ. Then in 1926, Houston's "favorite son" Jesse Jones, became sole owner of the *Houston Chronicle* and desired for his newspaper to enter broadcasting. In the month of March, it was announced that KFVI had been relocated to the Chamber of Commerce building on Milam Street, which the *Houston Chronicle* owned and maintained offices there. A soundproof studio was constructed on the building's fourth floor and the operating room was placed on the rooftop.[7]

After the move, the *Chronicle* "adopted" KFVI. On most days, the newspaper detailed the programming activities of the station and provided the news that would come from it.

KFVI produced some groundbreaking comedic formats that would define the "golden age of radio" of the 1930s and 1940s. Two of the station's first announcers can be credited with producing these innovative acts. They were Ted Hills and Jacob Schwarz. Ted Hills was born around 1891 and studied engineering at the University of Washington. Later, Hills entered World War I as a naval air serviceman where he learned to fly airplanes and was trained in radio communication. Hills' interest in wireless continued into his civilian life as he experimented with vacuum tubes and applied them to radiotelephony transmissions in an early post-World War I era. After World War I, the war department detailed him to Ellington Field where he fell in love with the Houston scene. Following his duty at Ellington, Hills remained in Houston and conducted broadcasting experiments over his amateur station. He was employed at the James Furniture Company as head of its radio department where he built and sold radio sets. Developing a reputation around the city as a skilled radioman, Hills was approached with the assignment of operating KFVI and serving as its announcer in 1925.[8]

Jacob Schwarz was the other announcer who seems to have had an influence on the entertainment presented over KFVI. Nicknamed "Dad," among other nicknames he would acquire over the years, Schwarz arrived in Houston around 1912 at the age of 48 as an entertainer connected to vaudeville acts. During the 1910s, he managed movie/vaudeville and cabaret theaters as well as a booking agency for stage performers. In his later years and nicknamed, "Grandfather," Schwarz would be associated with Will Horwitz's entertainment venues. Recognizing radio broadcasting as the next entertainment forum for live acts, "Dad" Schwarz accepted the opportunity to appear and book acts for KFVI.[9]

Hills and Schwarz seem to have teamed up to become the short-lived radio act of "Pops and Ted" at KFVI. "Ted," was of course, Mr. Hills while "Pops" was another of Schwarz's nicknames acquired and twenty-seven years older than Hills when they began appearing regularly over KFVI in 1926. Serving as the station's announcers, the comedic duo would go on before and between the musical acts to provide some light-hearted jokes, frolic and "pep" talk. Often "Pops and Ted" would close out the evening's programs with some departing comedic skits.[10] Here were the origins of the comic radio personality. The golden age of radio acts such as *Abbott & Costello, Amos n' Andy* and even today's comedic personalities can trace their beginnings to acts like "Pops and Ted" that were cracking radio jokes as far back as 1926.

Hills and Schwarz also staged some of Houston's most offbeat, light-hearted radio programs at the time. On March 23, 1926, KFVI aired "the second meeting" of the "Royal Order of the Flea." A tongue-in-cheek parody on fraternal groups that were popular at the time, the "Order of the Flea" was organized as a "goodwill suit for listeners." For this show, listeners who called to request songs from the radio act were made honorary members of the "Order" and were invited to visit the studios to perform a number or contribute to a charity of their choice.[11]

Another "voice" making a mark for itself at this time on KFVI was Kern Tips. Born in 1904 in Houston, Tips attended the Rice Institute in 1923 as WRAA was first airing the university's football games. Tips may have assisted in these broadcasts as he was assigned as a cub-sports reporter for the *Houston Post* and later, the *Houston Chronicle* from 1924 to 1926. When the *Chronicle* involved itself with KFVI in 1926, Tips served as an on-air sportscaster, giving details on the local sports events. By 1930, he was editing a daily sports column in Houston and was a popular commentator on local radio. Tips was remembered as a fast speaker while on the air. Walter Cronkite, the nationally renowned 20th century reporter who worked in the Houston market in the 1930s recalled that Tips would paste his pages together into a continuous scroll to keep up with his fast-paced announcing. For decades, Tips would be known as the "voice of the Southwest Conference" as he presented the action of Texas collegiate sports through shortwave radio to the world.[12]

The earliest regularly aired commercials in Houston can also be traced to KFVI. A dealer of automobiles, the Stodder Motor Company, sponsored hour-long weekly programs from the station, as did the Goodell Piano Company, a leading dealer of musical instruments, in June of 1926.[13]

KFVI remained under the ownership of the 56th Cavalry Brigade until early 1927. Then, like so many entities owning broadcasting stations, the militia unit discontinued management of it. With months remaining on its license, the local branch of the Atwater Radio Company assumed KFVI. A nationwide leader in manufacturing radios, the Atwater Radio Company had recently established an office in Houston to build household radio receivers. The acquisition of KFVI enabled the company to broadcast, test and advertise its products in a small market of broadcasting stations. The "new" station debuted from 7:30 p.m. to 10:00 p.m. on May 4, 1927, with a speech from the company's president, Frank Atwater, who came down to unveil the new broadcasting studio facility on the company's grounds. It remained at 239.9 meters and maintained the friendly policy of not airing at times when KPRC was on. Among those who operated the station at the radio-manufacturing plant during its brief existence were Gerald Chinski and Howard Van Eaton, both of WEAY fame.[14]

By 1928, KFVI had faded into broadcasting history and almost into complete obscurity. Some of the persons associated with the station, like Jacob "Pops" Schwarz, probably did not make regular appearances on Houston radio anymore, while others would go on to tinker with broadcasting as they "Kept Fooling Vid' It."

William John Uhalt reached the zenith of his career as a radio broadcaster on December 23, 1929, when his upgraded KTUE went on the air from Houston's Texas State Hotel. Started as a small 5-watt station in 1926, KTUE graduated into a 100-watt unit and was uniquely located in Houston's newly constructed luxurious 19-floor hotel at Rusk Street and Fannin.[15]

During his years employed with the federal government in the early 1920s, Uhalt supervised the installations of elaborate underground radio facilities, a design he would eventually carry over and realize with his KTUE station in 1929. A first for this region and for Houston hotels, the "new" KTUE's facilities were located underground in the basement of the Texas State Hotel. The station's reception offices, announcer's booth and two studios were situated under Houston's bustling downtown streets to create as much a soundproof environment as possible. "By going below the surface of the earth with the microphone, we have discovered that traffic and other noises are best kept from the transmitter to be broadcasted to listeners,"[16] said Uhalt on his design. KTUE's transmitter remained on top of the *Chronicle* building blocks away.

The station's initial three and a half-hour debut from the hotel included speeches from the city's mayor and pastors from Houston's leading denominations. Music was provided by the two jazz orchestras of Julien Paul Blitz and Paul Berge, past radio performers in Houston.[17] Subsequent daily programs from the station would often start at 9:00 a.m. and cease around 11:00 p.m. KTUE frequently aired afternoon remotes from locales such as the Rice Hotel and the Lamar Hotel, a throwback to WEAY ambitious "remotes" and a practice that KPRC was steadily abandoning by the late 1920s. Local advertisers such as the Thomas Goggan & Brothers Music Store and artists flooded the airwaves of KTUE.[18] The 100-watt KTUE promoted the Texas State Hotel and invited listeners to visit its entertainment facilities.

It can be said that no other Houston broadcaster had such an exciting and prestigious life as an amateur operator than William John Uhalt. A native of New Orleans, the young Uhalt learned about radio in 1907 at the age of sixteen amid New Orleans' hotbed activity in wireless. Before 1910, the Louisiana coastal city had several powerful wireless stations, exposing the young Uhalt to radiotelegraphy. A family legend claimed that the young Uhalt became so proficient in wireless that in

1912 he was able to pick up the faint distress signals of the ill-fated *Titanic* steamship sinking hundreds of miles away. As a teenager, he served part-time as a "Sparks" on merchant ships until the eve of World War I. During the war, Uhalt was eventually promoted to first-class electrician and chief radio electrician. His service in the war included covert radio operations.[19]

For six months during the war, he was stationed in Port Arthur, Texas where he would intercept hidden radio messages between Germany and Mexico. His deciphered messages were sent daily to Washington, D.C. After the war, the United States toured a captured German submarine, the *UB88*, along the Gulf Coast. The navy commissioned Uhalt to inspect the radio equipment used in the submarine. Uhalt also became the district's radio inspector and traveled extensively along the lower Gulf Coast to supervise the installation of government stations.[20]

After the war and in New Orleans, he and his brother, Joseph H.Uhalt, opened a chain of three electric stores. Not to be left behind during the broadcasting boom in the early 1920s, the Uhalt brothers installed WCBE, a small New Orleans station. Meanwhile, William Uhalt continued to serve the federal government as a radio inspector for its installations. Traveling through Texas on one of his radio-related duties, Uhalt met and eventually married a Texas woman. To satisfy his wife's desires in remaining near her family, he sold his interests in the electric stores back to his brother and settled in Houston. Upon his establishment here he was contracted to install the 56th Cavalry Brigade's 10-watt KFVI and patterned it after his WCBE in design.[21]

Amid the scores of radio dealers in Houston, Uhalt opened the Uhalt Electric Company in 1925 with a location at 614 Fannin Street. A year later, in August of 1926, he was issued a license to broadcast with a 5-watt station. Granting his desire to have vanity call letters, Uhalt's station was designated as KTUE. When Uhalt signed on for a broadcast, he reminded his audience that for its radio needs it should "Kome To Uhalt Electric." The 1926 secondary slogan used by KTUE would remind the audience that Uhalt had been "nineteen years in radio" and an expert in wireless technology.[22]

Similar to WEV years before, the Uhalt Electric Company's station was located on the store's grounds with a makeshift studio ready to accommodate the artist and the transmitter on top of the building. Limited programming aired on most weekdays and for a few hours only, depending on KPRC and KFVI's schedules for the day. Assisting Uhalt at this time was Robert McKay, a radio salesman for the store and KTUE's program director. Young radio operator Edward Bourdon found part-time work at the Uhalt Electric Company and assisted with the station's operations as well. Aside from his radio operations, Bourdon would later be better known as a noted photographer for *Life Magazine* as he captured on film some of Houston's important events and scenes of the twentieth century.[23]

Through periodic communication with each other, the Uhalt brothers exchanged ideas, and incredibly, the directions they took with their respective broadcasting stations often paralleled each other. In 1927, when Joseph Uhalt's WCBE in New Orleans began airing broadcasts in relation to the newspaper, the *New Orleans Statesman,* William Uhalt also did the same in Houston. While KFVI ceased to broadcast, Jesse Jones desired his *Houston Chronicle* to maintain a broadcasting unit. The newspaper acquired an interest to KTUE and moved its transmitter from the electric store to the roof of the *Chronicle* building where KFVI had functioned the year before. WCBE was re-licensed to become the 1,000-watt WDSU and functioned from New Orleans' DeSoto Hotel. Then, Jones had William Uhalt transfer KTUE to the Texas State Hotel in 1929.[24]

Among the innovative programming featured from Uhalt's KTUE in late 1929 was the half-hour radio drama, *Mary Mason,* a prototype to the modern soap opera. Developed by KTUE's program directors, *Mary Mason* followed the daily adventures of career-minded Mrs.

Mason and her relations with her husband, mother and housekeeper. Local thespians provided the voices for the daytime radio drama that often delved into housekeeping tips and cooking recipes for the listening Houston housewife. To encourage listener participation, program directors invited the radio audience to submit food recipes for upcoming airings.

Several major developments occurred at KPRC from 1926 to 1929 as the station sought to expand its operations. Early in 1926, the station's studios were moved from the *Post's* publishing plant on Dowling Street at Polk to the new *Post* building on Fannin Street in downtown. Called the "skyline studios," KPRC's new showcase facility was a penthouse on top of the building's twenty-second floor. Surrounded by windows, the "skyline studios" provided an excellent panoramic view of Houston in every direction on a clear day. The station's transmitter and operating facilities remained at the publishing plant.[25]

In 1927, KPRC began participating in what was called "chain broadcasting." For some years now, certain stations "linked" with others through telephone lines to air the same programming. In 1923, Dallas' WFAA was connected to New York City's WEAF to air President Calvin Coolidge's speech to Congress on December 6.[26] As discussed earlier in this history, Houston's WEAY did the same in 1924 when it leased a connection line to air the Firpo-Wills Heavyweight Title bout.

Beaumont's KFDM created Texas' first broadcasting network called the Magnolene Chain when it began linking to other stations around the state in early 1927. At 7:30 p.m. on October 3, 1927, KPRC connected with the Magnolene Chain and other Texas stations including KFDM and Fort Worth's WBAP in celebration of that Fort Worth station's fifth anniversary. For this chain broadcast, KPRC artists from the skyline studios aired several numbers.[27] Later that year on Christmas Eve, the Magnolene Chain network went international and aired a program from Nuevo Laredo, Mexico. From a temporary studio in the Caballo Blanco Inn located in the border Mexican town, artists from Mexico went over the airwaves through Texas stations: KPRC, KFDM, WRR, WFAA, KRLD, WBAP, and WJAD.[28]

A year before Texas' Magnolene Chain was created, another national network was born that would significantly contribute to KPRC's programming. In 1926, three corporations concerned with radio and electrical applications purchased New York City's two pioneering chain broadcasters, WEAF and WJZ, to form a radio network called the National Broadcasting Company (NBC).[29] Months after KPRC was first linked to the Magnolene Chain, a decision was made to receive programming from NBC.

On January 4, 1928, KPRC aired NBC's *Victory Hour* sponsored by the Dodge Brothers Motor Company. Utilizing communication lines owned by the American Telephone & Telegraph (AT&T), the Houston station was connected to WEAF, designated by NBC as the "red" network. On that evening, WEAF received "feeds" from stations in Hollywood, Chicago, New York and New Orleans that Houstonians heard clearly over KPRC as never before.[30]

The following evening, KPRC received programming from NBC's "blue" network out of WJZ and KFDM's Magnolene Chain. The *Maxwell House Hour* at 8:00 p.m. (Houston time) presented orchestra music from New York through station WJZ. A simple switch of circuits at KPRC, and an hour later, the Magnolene Chain went beyond state lines and transmitted a musical group from Parsons, Kansas. In time, network broadcasts on KPRC became more frequent, often covering an entire night of programs.[31]

One of the more popular programs from NBC and carried by KPRC was *Amos 'n' Andy*, a comedy radio skit which premiered in 1928 about the misadventures of two black men. Actually portrayed by two white men, *Amos 'n' Andy* played up controversial African-American stereotypes. The weeknight half-hour program was complemented with a daily syndicated comic strip that was carried by many newspapers around the nation, including the *Houston Post-Dispatch*.

Affiliating itself with NBC meant that KPRC would have to upgrade some of its equipment. Just weeks before the National Democratic Convention met in Houston in late June 1928 to nominate Al Smith as their United States presidential candidate, the *Houston Post-Dispatch* proclaimed the "new" KPRC, now an affiliate of NBC. In actuality, the only change was its transmitter. It probably was no coincidence that since the Radio Corporation of America (RCA) was the majority owner of NBC that KPRC's old transmitter was exchanged for a more powerful RCA model. During daytime hours, the transmitter was activated for 2,500 watts. In the evening, as radio signals required less effort to travel farther, its power was reduced to 1,000 watts.[32]

In that same year, for improved transmitting ability, KPRC moved its transmitter from the publishing plant to a remote area outside of Houston. Plans were in the works to install a 5,000-watt transmitter, by far the largest in Houston's market. Such a powerful transmitter meant having a huge facility available to house its generators and workroom, a luxury not feasible in downtown Houston. Land was acquired miles southwest outside of the city limits in the community of Sugarland to house the transmitter's facilities. Less than a mile away from the main sugar-manufacturing plants that made Sugarland what it was known for was KPRC's two-story complex with two towers, surrounded by a barbed wire fence. On November 15, 1929, the new transmitter for the station officially went on the air generating 5,000 watts of power.[33]

Although much of KPRC's evening programming came from networks in the late 1920s, program director Alfred Daniel managed to still provide locally based endearing entertainment. Starting in 1926, KPRC began a Houston tradition that decades later would continue.

Every Sunday morning broadcast would commence with airing the beautiful sounding chimes ringing from the belfry of the Saint Paul's Methodist Church in Houston. On one of these occasions, listeners called to the station's control room to report hearing a chirping bird among the ringing chimes. Manning the operations at the church that morning was Alfred Daniel, who climbed into the belfry to find a mockingbird singing its own song over KPRC.[34]

In 1927, the station ran a series that can be considered a predecessor to the genre of the radio and television quiz show. For many weekday nights, when not interrupted by network broadcasts, *What Do You Know About Your Country* quizzed listeners with trivia questions pertaining to the nation's geography, history or culture. Answers to these questions appeared the following day in the *Houston Post-Dispatch*.[35]

In 1929, KPRC moved its transmitter out to Sugarland, Texas. *Courtesy of Houston Post-Dispatch archives, Houston Public Library*

Houston Buffaloes baseball games continued to be broadcasted as KPRC technicians traveled with the team. Home games from Buffalo stadium in southeast Houston also were aired. Local businesses like the Star Electric and Engineering Company and the Levy Brothers Dry Goods Company sponsored the hour-long programs.[36]

Along with Frank Tilton and Judd Mortimer Lewis, other radio personalities emerged from KPRC. Starting with a talk given behind the microphone from station WEAY in 1923, Dr. A. H. Flickwir became a regular on Houston radio until 1929.

Educated as a physician and serving as the city health officer since 1920, Dr. Flickwir often gave weekly talks on KPRC concerning public health issues.[37] Afternoon musicals from Will Horwitz's theater, the Texan, were common as well. During these years the theater's pipe organist, Frances R. Clark, frequently was noted on the daily schedules to give recitals. On the music scene was "Tex" Lynn, a young man with an excellent range as a

Interior of KPRC's transmitter facility in Sugarland, Texas. *Courtesy of Houston Post-Dispatch archives, Houston Public Library*

whistler who was invited to return again and again to whistle blues and folk tunes.[38]

One of KPRC's longest running shows in the twentieth century Houston radio premiered on January 6, 1929. On that Sunday afternoon the KPRC skyline studios were transformed into a chapel. Known as the *Church of the Air,* the weekly program featured local ministers and inspirational music from various denominations. William States Jacobs of the First Presbyterian Church conducted the opening prayer followed by live music under the direction of Charles C. Hard. Reverend Hard would go on to minister and direct the *Church of the Air* for more than 20 years on KPRC.[39] At its debut, a prayer was invoked for the religious program and KPRC's longevity.

The prayer was answered.

Chapter Twelve

Not since 1922 did Houston experience such an emergence of new broadcasting stations then in 1930. Apparently, the major national stock market crash of 1929 did little to dissuade Houstonians from entering broadcasting. Three stations, all with sizable financial backing, initiated broadcasting operations in 1930. Before 1930, the last Houston broadcasting station to be licensed was KTUE in 1926. This resurgence can be attributed to two corporate bodies competing for what they believed would soon be a thriving Houston radio market.

One person unsure about the radio scene in Houston and how the recent stock market crash would affect him was Harvey Wheeler. A KPRC technician since 1928, Wheeler left the station in early 1929 to work at a factory in California manufacturing radio receivers. Recent downturns in the nation's economy convinced him that he should return closer to Houston. A phone call to his former boss at KPRC, G. "Eddie" Zimmerman, reaffirmed his decision to come back to Houston. "We're getting ready to put up *another* station here. We got a job for you. Come on,"[1] Zimmerman told him. What Wheeler discovered during that phone conversation was that components of the *Houston Post-Dispatch* were planning a second broadcasting station.

Success stories such as NBC's persuaded many around the nation to follow suit and seek to generate a market in broadcasting. By the late 1920s it had become clear that radio would provide new financial opportunities to those that owned and operated broadcasting entities. Huge revenues generated by radio advertising and by exploring the possibilities of creating a successful network or linking service as exemplified by Texas' Magnolene Chain were anticipated.

Associates of the *Houston Post-Dispatch* conceived of such a corporation that not only would have an interest in KPRC but would also aggressively seek to capture a foothold in Houston's broadcasting market by creating other radio stations. A charter was filed with the State of Texas in July of 1929 to create the Houston Broadcasting Company. Initially, three businessmen, L. V. Echols, Wilbur M. Cleaves and Mark Hopkins served as the corporation's stockholders as well as its officers.[2]

As specified by its charter, the Houston Broadcasting Company's purpose was to "engage in radio telegraphy and telephony, wireless telegraphy and telephony, with authority to own, lease, conduct, maintain and operate all the necessary plants, equipment and facilities..." for fifty years. Divided into 100 shares, $1,000 in capital stock was created with L. V. Echols as the majority shareholder owning forty percent of the stock.[3]

When initiating its venture into broadcasting, the Houston Broadcasting Company went the most economic route it could. KGHX was a station licensed in nearby Fort Bend County and was owned by the school system of that county. By late 1929, KGHX was not broadcasting anymore. The Houston Broadcasting Company acquired the station's license and its abandoned equipment was transferred to Houston. To facilitate the move, the Houston Broadcasting Company had financed KPRC's new 5,000-watt transmitter and moved it to Sugarland. Offices for the new broadcasting company were located in the new *Post* Building where KPRC's skyline studios sat atop the twenty-second floor.[4]

KPRC's old 1,000-watt transmitter atop the newspaper's publishing plant remained at that site but was downsized to 100 watts for the new station. Returning to the heyday of WEAY, the new station had its studios installed on the fourth floor in the old *Post* building, exactly the same location where Roy Garrett Watson's *Houston Post* cooperated with Will Horwitz's broadcasting venture of 1922. With everything set, it was announced that the station would sign in early 1930 with the call letters KTLC.[5]

A day before KTLC was to debut; the *Houston Post-Dispatch* devoted an entire newspaper section to the new station. Detailed articles about the station gave readers just about everything they wanted to know about the new station.

Several persons associated with the station wanted to create a slogan using KTLC's call letters. *Post-Dispatch* managers cited "Know Texas Largest's City," while Judd Mortimer Lewis, early KPRC announcer known as "Uncle Judd" on the air, gave the call letters a more personal flavor when he chose "Katie Elsie," a name similarly pronounced like KTLC. In truth, though, no intentional slogan was probably thought of when KTLC was chosen.[6]

At 1,500 kilocycles, 199.9 meters (1500 kHz) on the radio band,[7] KTLC owners stressed that the station's primary purpose was to provide the Houston market with local talent and artists, a flipside to KPRC that had now become an affiliate of NBC.

KTLC's program director for its debut was Ted Hills. After leaving KFVI in the mid-1920s, Hills continued in broadcasting when he joined KPRC as an announcer in 1927 or 1928.[8] His creative talent for presenting comedic radio acts quickly made Ted Hills a popular KPRC personality. For KTLC's first show on February 3, 1930, Hills helped to create quite a unique program. The first show was a musical extravaganza involving comedy with some 150 performers airing their talents. "King Amuse Me," a local basso singer, was the centerpiece of KTLC's debut. Adorned in a crown, royal cape, sitting on a throne and holding a regal staff, "King Amuse Me" was appointed as judge to KTLC's debut performances. Radio acts paraded into KTLC's studios, also known as King Amuse Me's court, attempting to entertain the difficult monarch. Between musical acts, comic duo "Frank and Herman" known as Frank Colby and Herman Engle in private life stood behind the microphone and provided listeners descriptions with the on-going attempts to entertain the king as they themselves tried to make "King Amuse Me" at least smile with their comedy. Those who failed to meet the king's standards that evening were given the guillotine, in fact merely a humorous sound effect heard by radio listeners.[9]

In KTLC's second day of broadcasting, the series, *The Cobra* debuted and then aired on Monday evenings at 8:00 p.m. The half-hour mystery radio show was somewhat a departure in Houston at the time.[10] Whereas KTUE's *Mary Mason* provided radio drama geared towards women during weekday mornings, *The Cobra* may have been one of the first local attempts to bring intrigue and drama to evening Houston audiences. It was a short run series, and practically no documentation survived of its existence, as was the case in so many other aspects of Houston broadcasting at this time.

For years, Houston's most powerful citizen, Jesse Jones, had shown a relaxed interest with the city's broadcasting scene. When Jones acquired the *Houston Chronicle* in 1926, the newspaper had KFVI move its studios to nearby offices of the *Chronicle*. Following KFVI's demise, Jones made arrangements to have KTUE replace it. Then a year after Jones built the Texas State Hotel, KTUE moved its studios into the basement of the hotel.[11]

Now, in the late 1920s, family members and associates of Jones actively sought his organization to open another broadcasting station. In 1929, M. Tilford Jones managed Jones' successful lumber company and handled the Jones organization's involvement in broadcasting. His first duty as Jones' broadcasting manager had KTUE transferred over to the Texas State Hotel, and now it was time for Jesse Jones' other gem, the Rice Hotel, to have a station. Months before the new station went on the air, a license had been granted in late 1929 with the call letters KTRH.[12]

Years later, Edward Bourdon would recall how he claimed to be the first voice heard over KTRH. An early radio operator with station KTUE and subsequent noted photographer, Bourdon built radio sets and was in awe when he heard Houston's first broadcasts coming from WEV, WCAK and WEAY.

At the age of 20 in 1930, Bourdon had already gained a reputation as a skilled radioman when the Jones organization asked him to join its team of technicians to build KTRH. The young Bourdon was assigned to build an amplifier for the station, a detail he had done before for local churches. Initial work on KTRH began in the basement of Jesse Jones' Gulf Building where the station's components were being assembled for the final installation in the Rice Hotel.[13]

For KTRH's sixtieth anniversary in 1990, Bourdon still had his written scribbled notes he recited over the air on the Tuesday evening of March 25, 1930. "Testing, testing. This is KTRH ...making equipment tests. We'd appreciate hearing from anyone who might be receiving this signal." This call to all listening parties traveled miles of communication lines to the station's 1,000-watt transmitter located just southeast of Houston in the rural community of Deepwater, near Pasadena where M.Tilford Jones owned land in his Deepwater-Houston Land Company. Acres of strawberries, the area's main produce at the time, surrounded KTRH's transmitter and facility.[14]

Equipment initially used by KTRH was acquired from KUT, the University of Texas station that officially signed on in 1925 and was now off the air by 1929. The transmitting equipment was still marked as "Property of University of Texas" when it arrived at KTRH's Deepwater facility.[15]

Like KPRC, KTRH's technicians set up residence near the transmitter to be in constant readiness should problems arise. A relative to Jesse Jones, T. Frank Smith, served as KTRH's first chief engineer and maintained a home near the Deepwater transmitter. Smith was an avid amateur operator and electrician in Dallas in the 1920s before he became an influential figure in the Jones organization's decision to build KTRH.[16]

KTRH stood for "Kome To Rice Hotel," an on-air slogan promoting the hotel's facilities. Upon visiting the luxurious hotel, one would find the station's studios and offices on the sixth floor. Monitoring and instrument rooms where technicians were able to view the progress of the broadcast separated the station's two studios. Studio A was fitted for ensemble musical groups while Studio B served orchestras and small bands. Visitors of KTRH could view the broadcasts from behind curtained glass windows.[17]

With the frequency at 267 meters, 1,120 kilocycles (1120 kHz), KTRH's first official speaker was Jesse Jones on March 25, 1930. Jones was followed by a religious invocation; a moving speech detailing Texas' history and a musical lineup of Houston's most popular radio stars at the time. That evening, Jerry Sloan and Frank Threadgill, a duo billed as a "cowboy harmony team," performed hillbilly music with a Western swing, which was quickly becoming a popular style in Texas. Popularized at KPRC, pianist Frank Tilton and the young "Tex" Lynn who whistled out tunes were also on KTRH's first billing. The lineup included Paul Berge's Rice Hotel orchestra and local glee clubs.[18]

On that March 25 evening, KTRH also introduced itself as a local affiliate of the recently renamed Columbia Broadcasting System (CBS) by airing programs from this network. A merger of two radio networks in 1927 formed the Columbia Phonograph Broadcasting Company. After a year of struggling, a wealthy man enamored with radio bought the fledgling Columbia Phonograph Broadcasting Company and changed the network's name to the Columbia Broadcasting System in January of 1929.[19]

Like the agreement between KPRC and NBC, most "feeds" received from CBS to KTRH occurred for afternoon and evening programming. Now, in the spring of 1930, Houston had quite a diverse radio scene; two stations connected to major networks, a local group interested in turning broadcasting into profits called the Houston Broadcasting Company and another Houston-based effort gearing itself to compete in the city's radio market.

Perhaps no other Houston broadcaster suffered as many financial setbacks in the Depression years as William Uhalt did. In July of 1929, Uhalt lost his owning rights to the Uhalt Electric Company when he sold the store to other business parties. Months later, he reached the peak of his broadcasting career when M. Tilford Jones and proprietors of the Texas State Hotel agreed to upgrade and relocate his KTUE to the basement of their fine hotel.[20]

But when Uhalt agreed to use his KTUE as a promotional vehicle for the Texas State Hotel, it was apparent that broadcasting was becoming a big business enterprise with the independent broadcaster, like himself, being gradually phased out. In time, a local corporation seeking to compete with the Houston Broadcasting Company purchased KTUE's interests from Uhalt. In February 1930, three businessmen chartered the KTUE Broadcasting Company. Charter terms specified that the corporation would endure for "50 years" and "engage in radio telegraphy and telephony." Investments were divided up between three stockholders with Fred J. Heyne holding eighty percent of the stock. A close business associate of Jesse Jones, Heyne served as vice-president while Jones was president of the Houston Properties Company, a corporation dedicated to promoting the creation of skyscrapers in Houston.[21]

Five months later in July 1930, the KTUE Broadcasting Company rechartered and changed its title to the Harris County Broadcasting Company, since the KTUE license would be expiring at the end of the month. On July 31, 1930, KTUE aired its last program as the Harris County Broadcasting Company readied to transfer the equipment.[22]

The studios remained in the basement of the Texas State Hotel, but the management changed and the call letters were replaced with KXYZ. The station's transmitting tower was placed atop the city's Gulf Building, Houston's tallest edifice with thirty-two floors. Measuring 590 feet above ground, the tower on the building was a familiar landmark in Houston's skyline for many years.[23]

On October 20, 1930, KXYZ, billed as "everybody's station," debuted with little hoopla. Thurston May, formerly connected to KTRH and now manager of KXYZ, opened the station's premiere with the simple statement, "This is station KXYZ, Houston, Texas." Powered with 100 watts and on the frequency of 211.1 meters, 1,420 kilocycles (1420 kHz), the station followed May's very brief introduction with the Schott's Bakery sponsorship of a two-hour program of "favorite musics of the world." Artists offered renditions of popular songs in French, Spanish, German, Russian and Italian. Other local artists followed and performed until midnight that evening.[24]

Though it still had an unlisted telephone number during its first hours of broadcast, the station received approximately a hundred and ten phone calls to congratulate its debut. Later, the station received another 265 calls when it signed off that evening.[25]

An estimated 24,173 or close to thirty-two percent of Houston's households had radios to enjoy the city's four stations operating by the end of 1930.[26] Of course, these figures may not have reflected the number of amateur operators that were still out there devising their own radio sets out of common household items. Two years passed before KTLC would also sign off permanently. Tighter restrictions imposed by the Federal Radio Commission prevented a group from owning more than one station in a market, a regulation that heralded a premature termination to companies such as the Houston Broadcasting Company and the Harris County Broadcasting Company. In December of 1939, a motion was filed by the Houston Broadcasting Company to dissolve itself and cease business activities. The Harris County Broadcasting Company would survive until 1948 during which this time T. Frank Smith gained a majority ownership of the company. Local, flamboyant wildcatter Glen McCarthy and other stockholders purchased the company from Smith and the Jones organization to form the Shamrock Broadcasting Company.[27]

From McCarthy's newly constructed Shamrock Hotel, the Shamrock Broadcasting Company aired a popular weekly series called *Saturday Nights at the Shamrock* that aired programs nationally over KXYZ starting in 1948.[28]

Houston's radio scene before 1940 made other significant, enduring contributions to broadcasting. In 1932, KTRH began a series called *Man on the Street.* Microphones were installed on the street corners outside of the Rice Hotel where announcers Parks Johnson and Jerry Belcher conducted impromptu, brief surveys with Houstonians on how they would vote in the upcoming 1932 presidential election. The idea sometimes produced hilarious results and soon became a regular feature on KTRH. Later, trivia questions were posed to passersbys to add fun to the unplanned interviews with prizes awarded for correct answers (arguably, the origins of the quiz show genre). In 1935, the format was picked up by national networks and called *Vox Pop* (Voice of the People) where it ran until 1948.[29]

KTRH's transmitter in Deepwater was upgraded in 1936 to simultaneously beam out KXYZ's and KTRH's signals. Through a series of filters, both stations could transmit from the same antenna tower at the same time, reportedly only the second of its kind in the world during this time. For over forty years, KTRH maintained studios at the Rice Hotel until 1973 when it moved just miles southwest of downtown Houston.[30]

During the 1930s, KPRC, KTRH, and KXYZ served as the only stations in Houston. In fact, the assignments for these stations still exist as Houston entered the 21st century.

The people who played vital roles in shaping Houston's first twenty-six years of radio would go on to leave lasting impressions in the city's history. Some would make radio broadcasting a lifelong career and become notable Houston figures, while others would carry their knowledge of radio technology into other fields. Some would gravitate to the next mass medium, television.

After successfully operating station WEV, Ben James Still would sell his interests in the Hurlburt-Still Electric Company to his two other partners in 1926 and form his own business - the Ben Still Electric and Radio Company. Ben Still remained a popular figure in the business of selling radios but never maintained a broadcasting unit again. He passed away in 1945.[31]

Known as one of Houston earliest and most influential amateur operators, Louis Peine completed his stint at WEV to move on to other broadcasting stations such as KPRC. In later years, he worked with General Electric in designing radio equipment for the aviation industry. Reflective of his service during World War I, Peine also trained World War II servicemen in radio technology on the campus of the University of Houston. After decades in Houston, Louis Peine retired with his wife in Texas' lower Gulf coast area and passed away in 1969.[32]

Pre-broadcasting era radio operators Clifford Vick and Alfred Neal Dargan capitalized on their expertise in radio by opening business ventures. Hailed as a hero for receiving news dispatches from strickened areas during the hurricane of 1919, Vick became an early broadcaster and opened Vick's Radio Construction Company. He later abandoned his radio business to enter the oil industry, but always maintained his interests as an amateur operator. Alfred Neal Dargan and fellow Houston Radio Club member Hallet Worthington opened the Radio Den store in the 1920s. Dargan lived his entire life in Houston, held a high position in the local offices of General Electric and maintained a close lifelong friendship with Louis Peine.[33]

Gerald Chinski pursued broadcasting with a passion. His years with WEAY convinced him that he had found his calling. While he lived in Houston, Chinski served as technician to Houston's most popular stations including

KXYZ. In his retirement years, he moved to Michigan with his family and passed away in 1980. Chinski's best friend and assistant at WEAY, Frank Austin, took his radio experience in a different direction. After the demise of WEAY and after receiving more formal education in electrical engineering, Austin accepted a position with the Humble Oil and Refining Company to design radio equipment in the company's geophysical research area. He later was appointed as a vice-president for the Houston Lighting & Power Company.[34]

Others from Houston's pioneer broadcasting era would venture into southern Texas to find their fame in radio and television. After several years of serving as engineer for KTRH and KXYZ, T. Frank Smith and several Houston associates purchased radio station KGFI in Corpus Christi and changed its call letters to the more appropriate, KRIS in 1936 and later added a second station with KVAL in Brownsville. Meanwhile, Smith's partnership in the Harris County Broadcasting Company allowed him to diversify into the next broadcasting medium, television.

Following the selling of the Harris County Broadcasting Company in 1948, Smith settled in Corpus Christi and eventually created a television station KRIS-TV in the 1950s, that city's first TV station.

Will Horwitz continued with his ambitious and controversial ways. He eventually acquired Houston's Uptown and The Ritz movie theaters to add to the Iris and The Texan. During the 1920's, he collaborated with the *Houston Post-Dispatch* to start a popular citywide annual Christmas party for needy children providing gifts during the Depression years. Horwitz has also been credited to building one of the city's first pedestrian-friendly tunnel systems that connected two of his movie theaters with one air conditioning unit that cooled both theaters. In later years, Horwitz's controversial side would surface involving a new venture into radio broadcasting.[35]

In 1931, he bought a part-ownership in broadcasting station XED in the border town of Reynosa,

Mexico. XED was closely monitored by U.S. authorities for its lack of following U.S. regulations by transmitting at 100,000 watts, a strength that would severely interrupt transmitting, licensed U. S. stations just across the border. During Horwitz's ownership, XED announced Mexico's lottery, a gaming device that was illegal in the United States. Because of an illegality stemming from the lottery, Horwitz and four others, including his wife and former KTRH announcer Milton Hall, were arrested for smuggling lotto tickets across the border to his American friends.[36]

Found guilty, Horwitz was sentenced to eighteen months at the nation's hardened Leavenworth prison. But scores of appreciative Houstonians remembered the showman's generosity and repaid it with petitions and protests to reduce his time. It worked. Just six months into his prison term, Horwitz was released. President Franklin Delano Roosevelt later granted him a full pardon in 1940. A year later, in 1941, Horwitz, at the age of 55, suffered a massive heart attack and died on the morning of Christmas Day. In his "P. T. Barnum" fashion, Horwitz was credited with addressing the annual Christmas party from his deathbed with, "The show must go on."[37]

KPRC's first general manager G. E. Zimmerman traveled to several states around the nation to accept positions with NBC in New York City and later with a station in Arkansas. Following the sale of KTUE, William John Uhalt remained in the radio business in Houston and ultimately met an untimely death in 1936 at the age of 45.[38]

Among those who continued broadcasting careers in Houston for many years were Ingham S. Roberts III, Kern Tips and Ted Hills. Following his years with WGAB, Roberts became an announcer with KPRC until the late 1920s. He entered other radio ventures and was appointed as a manager of KPRC in the 1930s. Sports announcer Kern Tips held a huge radio following. His popular sports commentaries and announcing during the early 1930s galvanized his appointment as manager of KPRC from 1935 to 1946. For thirty-two years, he was known as the "voice of the Southwest Conference" as he brought Texas collegiate events to hundreds of radio fans.[39]

Tips' voice went beyond the nation's borders and was beamed around the world through the facilities of the Armed Forces Radio Service network. After retirement in 1966, Tips produced and narrated a radio-syndicated sports series featuring a five-minute show aired throughout the Southwest. In 1967, Kern Tips passed away in Houston at the age of 62.[40]

What started as a hobby for Ted Hills developed into a groundbreaking career. His positions at KFVI, KPRC and KTLC led to a decade at KTRH as a production manager during the 1930s. Then, he was offered the opportunity of a lifetime when entered the new forum known as television. While working at radio station KLEE in the late 1940s, Hills accepted a position in Houston's first television station, KLEE-TV, in 1949. KLEE-TV lasted only a year when it was purchased and transformed into KPRC-TV in 1950. Subsequently, Hills returned to radio and died in 1960.[41]

William Hobby also diversified from radio into television broadcasting. As publisher of the *Houston Post-Dispatch* in 1924, Hobby played an essential role in getting radio station KPRC on the air in 1925. The former Texas governor and philanthropist remained a constant advocate of Houston broadcasting in the first half of the twentieth century. As television broadcasting was being developed in the century's second half, Hobby took interest. He purchased the failed KLEE-TV and turned it into KPRC-TV in 1950, the city's first enduring television.[42]

Alfred Daniel would remain at radio station KPRC for thirty years. Houston's radio dean lived long enough to see the infancy of wireless telegraphy gestate into the principles of television when he passed away in 1955. His preference, of course, remained in radio broadcasting. And while comparing radio to television broadcasting, he argued this point: "But did you ever hear music from a crystal set? That's the sweetest music ever broadcast."

Chapter 1 Endnotes

[1] Mary Frazier, "A Life in Radio: By a Gypsy Fortune Teller," Houston Press 13 Jan. 1948: 5.
"Houston To Have Wireless Station," Houston Chronicle 29 May 1906: 5.

[2] "Political News Without Wires," Houston Chronicle 18 July 1906: 8.

[3] Susan J. Douglas, Inventing American Broadcasting (Baltimore, MD: The John Hopkins University Press, 1987) 92.

[4] "Houston To Have Wireless Station," Houston Chronicle 29 May 1906: 5.

[5] Lee de Forest, Father of Radio: The Autobiography of Lee de Forest (Chicago, IL: Wilcox and Follett, 1950) 218.

[6] "Conspiracy to Defraud Using Mail is Charge," San Antonio Express 2 Mar. 1911: 1.

[7] Susan J. Douglas, Inventing American Broadcasting (Baltimore, MD: The John Hopkins University Press, 1987) 93.

[8] IBID, 97.

[9] Tom Lewis, Empire of the Air: The Men Who Made Radio (New York: Edward Burlingham Books, 1991) 44.

[10] Erik Larson, Isaac's Storm: A Man, A Time and the Deadliest Hurricane in History (New York: Crown Publishers, 1999) 265.

[11] "Wireless Messages Sent," Dallas Morning News 7 Sept. 1904: 12.
Lee de Forest, Father of Radio: The Autobiography of Lee de Forest (Chicago, IL: Wilcox and Follett, 1950) 156-58.

[12] "Hurled Through Air," Dallas Morning News 4 Sept. 1904: 23.

[13] "Wireless Messages Sent," Dallas Morning News 7 Sept. 1904: 12.

[14] "Wireless Wires Out of Shape," Houston Chronicle and Herald 14 Sept. 1904: 1.

[15] "By Air Route to Houston," Houston Chronicle and Herald 16 Nov. 1904: 7.

[16] "Within Two Weeks: Station is All Ready," Galveston Daily News 15 Feb. 1905: 10.

[17] "The First Messages Was Forty Miles Out," Galveston Daily News 12 Mar. 1905: 24.

[18] Tom Lewis, Empire of the Air: The Men Who Made Radio (New York: Edward Burlingham Books, 1991) 49.

[19] Lee de Forest, Father of Radio: The Autobiography of Lee de Forest (Chicago, IL: Wilcox and Follett, 1950) 220.

[20] "Wireless Report of Legislature," Houston Chronicle 7 Mar. 1906: 1.
"Houston To Have Wireless Station," Houston Chronicle 29 May 1906: 5.
"Political News Without Wires," Houston Chronicle 18 July 1906: 8.

[21] "Wireless Opens: Messages Transmitted," Houston Daily Post 15 Aug. 1906:3.

[22] Frank Fayant, "The Wireless Telegraph Bubble," Success Magazine vol. 10, July 1907: 509.

[23] "Houston To Have Wireless Station," Houston Chronicle 29 May 1906: 5.
"Block for Wireless," Galveston Daily News 31 May 1906: 7.
"Land for Wireless: Poles to Come from Denver," Galveston Daily News 4 June 1906: 7.

[24] "A New Transformer," Galveston Daily News 20 Aug. 1906: 10.

[25] Diana J. Kleiner, "Brunner, Texas," The New Handbook of Texas, ed. Roy Tyler, vol. 6 (Austin, TX: State Historical Association, 1996) 243.

[26] 1906 city directory for Houston
"Land for Wireless," Galveston Daily News 4 June 1906: 7.
"Station Now is Complete: Ready for the Test," Houston Chronicle 12 Aug. 1906: 5.

[27] Curtis Bishop and L. R. Wilcox, "Telegraph Service," The New Handbook of Texas, ed. Roy Tyler, vol. 6 (Austin, TX: State Historical Association, 1996) 243.

[28] Marguerite Johnston, Houston: The Unknown City, 1836-1946 (College Station, TX: Texas A & M University Press, 1991) 59.

[29] IBID, 92.

[30] "Two Wireless Messages Go," Houston Chronicle 14 Aug. 1906: 11.
"Wireless Opens: Messages Transmitted," Houston Daily Post 15 Aug. 1906: 3.

[31] "Wireless at Work: Getting in Line Now," Austin Statesman 14 Aug. 1906.

[32] "Improve Galveston Wireless," Galveston Daily News 19 Aug. 1906: 12.
"A New Transformer," Galveston Daily News 20 Aug. 1906: 10.

[33] "Wireless for Port Arthur," Galveston Daily News 17 Aug. 1906: 6.
"Port Arthur Wireless is Working," Southwestern Electrician October 1906: 20.

[34] Frederick E. Emmons, American Passenger Ships: The Ocean Lines and Lines, 1873-1983 (Newark, NJ: University of Delaware, 1985) 113-115.
"The First Messages: Was Forty-Five Miles Out," Galveston Daily News 12 Mar. 1905: 24.
"Head of Wireless: Col. Wilson, Vice-president of De Forest Company, Here," Houston Daily Post 17 Oct. 1906: 11.

[35] Susan J. Douglas, Inventing American Broadcasting (Baltimore, MD: The John Hopkins University Press, 1987) 178.

[36] United States, Navy Department, Bureau of Equipment 1907. List of Wireless-Telegraph Stations of the World. by Thomas H. White. <http://earlyradiohistory.us/1907stat.htm.>

Chapter 2 Endnotes

[1] "Pioneer Radio Technician Alfred P. Daniel Dies Here," Houston Post 1 Feb. 1955: 1.

[2] Shannon Tompkins, "Echoes of the Storm," Houston Chronicle 8 Aug. 2000: 1.

[3] Marguerite Johnston, Houston: The Unknown City, 1836-1946 (College Station, TX: Texas A & M University Press, 1991) 186.

[4] IBID, 174-6.

[5] Susan J. Douglas, Inventing American Broadcasting (Baltimore, MD: The John Hopkins University Press, 1987) 190.

[6] "The Wireless is Now Here," Houston Chronicle and Herald 13 Aug. 1904: 5.

[7] IBID.

[8] Erik Barnouv, A Tower in Babel: A History of Broadcasting in the United States up to 1933 1st vol. (New York: Oxford University Press, 1966-1970) 28.

[9] Susan J. Douglas, Inventing American Broadcasting (Baltimore, MD: The John Hopkins University Press, 1987) 198.

[10] "Wireless By Kite," Southwestern Electrician March 1907: 23.

[11] IBID.

[12] IBID.

[13] Richard Schroeder, Texas Signs On: The Early Days of Radio and Television (College Station, TX: Texas A & M University Press, 1998) 19.
David McComb, Galveston: A History (Austin, TX: University of Texas Press, 1986) 158, 179.

[14] Erik Larson, Isaac's Storm: A Man, A Time and the Deadliest Hurricane in History (New York: Crown Publishers, 1999) 265.

[15] "Wireless By Kite," Southwestern Electrician March 1907: 23.

[16] Frank Fayant, “The Wireless Telegraph Bubble,” Success Magazine vol. 10, June & July 1907.

[17] United States. Navy Department. Bureau of Equipment 1909. List of Wireless Telegraph Stations of the World. by Thomas H. White. <http://earlyradiohistory.us/1909stat.htm.>
“Conspiracy to Defraud Using Mail is Charge,” San Antonio Express 2 Mar. 1911: 1.

[18] United States. Navy Department. Bureau of Equipment 1909. List of Wireless Telegraph Stations of the World. by Thomas H. White. <http://earlyradiohistory.us/1909stat.htm.>

[19] Susan J. Douglas, Inventing American Broadcasting (Baltimore, MD: The John Hopkins University Press, 1987) 169.

[20] “Uses Wireless Telephone: Lee DeForest Talks for a Distance of Two Miles at Norfolk,” Houston Chronicle 13 Oct. 1907: 32.

[21] “From Top of the Chronicle Building the Wireless Phone Will Send Message,” Houston Chronicle 26 Mar. 1909: 3.

[22] IBID.
“Wireless Phone for Houston,” Southwestern Electrician April 1909: 30.

[23] Tom Lewis, Empire of the Air: The Men Who Made Radio (New York: Edward Burlingham Books, 1991) 81.

[24] “De Forest is Coming to Install Texas Service,” Houston Chronicle 12 June 1910: 27.

[25] “Rice Hotel Has Wireless System,” Houston Chronicle 15 Jan. 1908: 11.

[26] IBID.
1908 city directory for Houston

[27] unpublished manuscript, History of the Texas Company’s Marine Department (Texaco, 1952) 13 in author’s possession.

[28] Bascom N. Timmons, Jesse H. Jones: The Man and the Statesman (Houston, TX: Owl Publications, 1956) 77.

[29] Dr. S.O. Young, A Thumbnail History of the City of Houston, TX. From Its Founding in 1836 to the Year 1912 (Knoxville, TN: H. W. Crew & Co., 1912) 174.

B. H. Carroll, Jr., Standard History of Houston, Texas, from a Study of the Original Source (Knoxville, TN: H. W. Crew & Co., 1912) 259.

[30] B. H. Carroll, Jr., Standard History of Houston, Texas, from a Study of the Original Source (Knoxville, TN: H. W. Crew & Co., 1912) 259.

[31] "Texas Company's Wireless Tower," Houston Chronicle 9 Oct. 1910: 15.

[32] advertisement in Houston Chronicle on July 10, 1910: 5

[33] Wallace O. Chariton, The Great Texas Airship Mystery (Plano, TX: Woodware Publishing, 1991) 59.

[34] Tom Lewis, Empire of the Air: The Men Who Made Radio (New York: Edward Burlingham Books, 1991) 104-07.

[35] advertisement in Houston Chronicle on August 14, 1910.

Chapter 3 Endnotes

[1] "Wireless Talk Tonight," Houston Chronicle 31 Jan. 1911: 6.

[2] "Talk on Wireless: Introductive Lecture at Y.M.C.A. by Practical Man," Houston Daily Post 22 Jan. 1911: 12.
"He Has Come to Arrange for Installation of Wireless Telegraph Station at A.M College," Bryan-College Station Eagle 27 Ap. 1911.

[3] "Conspiracy to Defraud Using Mail is Charge," San Antonio Express 2 Mar. 1911: 1.

[4] B. H. Carroll, Jr., Standard History of Houston, Texas From a Study of the Original Source (Knoxville, TN: H.W. Crew & Co., 1912) 260.

[5] "The Wireless in Texas," Southwestern Electrician July 1910: 17.

[6] "Wireless Hearing Ends," San Antonio Express 9 Mar. 1911: 1.
"Conspiracy to Defraud Using Mail is Charge," San Antonio Express 2 Mar. 1911: 1.

[7] B. H. Carroll, Jr., Standard History of Houston, Texas From a Study of the Original Source (Knoxville, TN: H. W. Crew & Co., 1912) 260.

[8] "Conspiracy to Defraud Using Mail is Charge," San Antonio Express 2 Mar. 1911: 1.

[9] classified advertisement in Houston Daily Post on April 17, 1910.

[10] "The Wireless Case: Preliminary Hearing of Dickinson-Beauchamp." Houston Daily Post 9 Mar. 1911.

[11] "The Wireless in Texas," Southwestern Electrician July 1910: 17.

[12] "Conspiracy to Defraud Using Mail is Charge," San Antonio Express 2 Mar. 1911: 1.

[13] "Wireless Telephone in Chronicle Building," Houston Chronicle 15 Sept. 1910: 11.

[14] IBID.

[15] "'Acquaintance Day'" Celebration a Success," Victoria Daily Advocate City Briefs section of 6 Aug. 1910: 3
[16] "The Wireless Case: Preliminary Hearing of Dickinson-Beauchamp," Houston Daily Post 9 Mar. 1911: 5.

[17] "Conspiracy to Defraud Using Mail is Charge," San Antonio Express 2 Mar. 1911: 1.

[18] "Texas Wireless in Commercial Operation," Southwestern Electrician Jan. 1911: 29.

[19] Bascom N. Timmons, Jesse Jones: The Man and the Statesman (Houston, TX: Owl Publications, 1956) 79.

[20] "The Wireless in Texas," Southwestern Electrician July 1910: 17.

[21] "Wireless Development: Texas Coast Country Soon to Have Wireless," Houston Chronicle 29 Jan. 1911: 32.

[22] "Wireless Working: City Electrician of Waco Sends Greetings to Houston," Houston Chronicle 20 Feb. 1911: 7.

[23] "Wireless to New York: The Texas System to Build a Powerful Relay Station Here," Houston Chronicle 5 Feb. 1911: 14.

[24] "Texas Wireless Officials in Trouble," Southwestern Electrician Mar. 1911: 28.

"Wireless Man Here: J.B. Dickinson, Financial Agent for the Texas Wireless System Explains," Houston Chronicle 1 Mar. 1911: 5.

"Wireless Station: In This City is Now Ready to Receive Messages," Victoria Daily Advocate 3 May 1911: 1.

[25] "Wireless Service," Victoria Daily Advocate 6 May 1911: 1.

[26] "Wireless is Used in Produce Business," Houston Chronicle 8 May 1911: 8.

[27] B.H. Carroll, Jr., Standard History of Houston, Texas From a Study of the Original Source (Knoxville, TN: H.W. Crew & Co., 1912) 260.

[28] "Wireless Men Found Guilty," Southwestern Electrician February 1913: 30.

Chapter 4 Endnotes

[1] Susan J. Douglas, Inventing American Broadcasting (Baltimore, MD: The John Hopkins University Press, 1987) 220.

[2] IBID.

[3] "Wireless Users Given Warning," Houston Chronicle 22 Jan. 1915: 9.

[4] Frank Matejka, "The First Football Game: Texas A&M vs. Texas University, November 24, 1921," unpublished manuscript, February 1980: 5.

[5] "Houston Amateurs Held Wireless Talks with Distant Friends," Houston Chronicle 3 Dec. 1916: 26.
1916 city directory for Houston.

[6] "Uncle Sam's Navy on Exhibit," Houston Daily Post 10 Nov. 1912.
"Factory Products Form an Interesting Exhibit," Houston Chronicle 10 Nov. 1912: 16.

[7] "Lecture on Electricity," Houston Chronicle 7 Jan. 1915: 9.

[8] IBID.

[9] "Houston Amateurs Held Wireless Talks with Distant Friends," Houston Chronicle 3 Dec. 1916: 26.

[10] United States, Department of Commerce, Bureau of Equipment, Radio Stations of the United States (Washington, D. C.: Government Printing Office, 1916) 122.
1915 city directory for Houston.

[11] Interview with Walter Peine on July 18, 1996.
"Wolly Peine," New Encyclopedia of Texas, comp. and ed. by Ellis A. Davis and Edwin H. Grobe (Dallas, TX: Texas Development Bureau, 1925) 1490.

[12] Interview with Walter Peine on July 18, 1996.

[13] Andrew Forest Muir, article on the Autry family found in The Flyleaf Oct. 1960: 5.

[14] Marquis James, History of Texaco: The First Fifty Years (Texas Company, 1953) 114.
[15] Letter from Commandant C.H. Druse to Battalion Adjutant Walton Jackson, February 8, 1912. James Lockhart Autry II Collection (Woodson Research Center, Rice University: Houston, TX.)

[16] Letter from James Lockhart Autry II to Junior Plattsburg Military Academy, May 7, 1918. James Lockhart Autry II Collection, (Woodson Research Center, Rice University: Houston, TX.)
United States, Department of Commerce, Bureau of Equipment, Radio Stations of the United States in 1916 (Washington, D. C.: Government Printing Office) 122.

[17] "Who's Who in Amateur Wireless," QST Dec. 1919: 23.

[18] Letter from James Lockhart Autry II to Junior Plattsburg Military Academy, May 7, 1918. James Lockhart Autry II Collection, (Woodson Research Center, Rice University: Houston, TX.)

[19] "Storm Blows into Prominence Amateur Radio Wizard Who 'Scooped' All Weather Experts," Houston Chronicle 21 Sept. 1919: 11.

[20] IBID.

[21] "Houston Boys in Radio Club Talk to Chicago," Houston Chronicle 12 Nov. 1916: 20.
"Who's Who in Radioland," Houston Chronicle. Dec. 1925.

[22] Susan J. Douglas, Inventing American Broadcasting (Baltimore, MD: The John Hopkins University Press, 1987) 205-06.

[23] "Wireless Demonstration is Feature of Electrical Week," Houston Daily Post 12 Nov. 1916: 23.

[24] Interview with James William Weatherford, Jr. September 25, 1996.

[25] "Wireless Demonstration is Feature of Electrical Week," Houston Daily Post 12 Nov. 1916: 23.
"Houston Boys in Radio Club Talk to Chicago," Houston Chronicle 12 Nov. 1916: 20.

[26] "Hurlburt-Still also Celebrating 20th Birthday," Houston Chronicle 16 Oct. 1921: 23.

[27] Interview with William S. Baker on June 6, 1996.

[28] "Company is First Local Radio Firm," Houston Post-Dispatch 19 Sept. 1926: 2, Radio Exposition section.

[29] Captain L.S. Howeth, USN. History of Communications–Electronics in the United States Navy (Washington, DC: Government Printing Office, 1963) 526.

[30] "Storm Blows into Prominence Amateur Radio Wizard Who 'Scooped' All Weather Experts," Houston Chronicle 21 Sept. 1919: 11.

[31] Interview with Leon Peine on June 27, 1996.

[32] Donald L. Bartlett and James B. Steele, Empire: The Life, Legend and Madness of Howard Hughes 1st edition (New York: Norton Press, 1979) 35, 44.
Pat H. Broeske, Howard Hughes: The Untold Story (New York: Penguin Books, 1996) 13.

Chapter 5 Endnotes

[1] Marguerite Johnston, Houston: The Unknown City, 1836-1946 (College Station, TX: Texas A&M University Press 1994) 196.

[2] Erik Barnouv, A Tower in Babel: The History of Broadcasting to 1933 (Oxford Unversity Press: New York, NY 1966) 292.

[3] Letter from James Lockhart Autry II to special agent W.W. Green, June 14, 1917. James Lockhart Autry II Collection (Woodson Research Center, Rice University: Houston, TX.)

[4] IBID.

[5] Letter from James Lockhart Autry II to the Junior Plattsburg Academy on May 7, 1918. James Lockhart Autry II Collection (Woodson Research Center, Rice University: Houston, TX.)

[6] Interview with Mrs. Billy Kilgore on December 12, 2002.

[7] Mrs. W. M. Baines, Houston's Part in the World War. (Houston, TX.) 175+.
Minutes from the Houston School System board meeting of November 12, 1917. (Houston Independent School District Administration Building: Houston, TX.) 1

[8] "87 Men Register for Radio Work," Houston Chronicle 14 Nov. 1917: 16.
"Two Classes in Radiotelegraphy to Begin Training," Houston Chronicle 19 Nov. 1917: 11.

[9] Interview with Walter Peine on July 18, 1996.
"Two Classes in Radiotelegraphy to Begin Training," Houston Chronicle 19 Nov. 1917: 11.
"Student Radio Classes Apt in Telegraphy Game," Houston Chronicle 27 Nov. 1917: 8.

[10] "City News in Brief" section, Houston Chronicle 13 Dec. 1917.

[11] "War Courses Have Speedy Progress: Practical Work Now Being Given in Gas Engines and Wireless Courses," Rice Thresher 2 Mar. 1917.

[12] Interview with James William Weatherford II on Sept. 25, 1996.

[13] Yearbook for Ellington Air Base of 1918, (Houston, TX: 1918) 85-86.
Art Leatherwood, "Ellington Field." New Handbook of Texas vol. 2, pages 827-8.
Claudia Hazlewood, "Camp Logan." New Handbook of Texas vol.1, page 941.

[14] Yearbook for Ellington Air Base of 1918 (Houston, TX: 1918) 85-86.

[15] Letter from James Lockhart II to Junior Plattsburg Military Academy on May 7, 1918. James Lockhart Autry II Collection (Woodson Research Center, Rice University: Houston, TX.)

[16] "Pioneer Radio Technician Alfred P. Daniel Dies Here," Houston Post 1 Feb. 1955: 1.

[17] Personal military papers of Alfred Patrick Daniel, service number 1088826. National Personnel Records, St. Louis, MO.

[18] Walter H. Johnson, "Advanced Radio School," Radio 15 Nov. 1918: 4.

[19] Yearbook for Ellington Air Base of 1918, (Houston, TX: 1918) 87,161.

[20] Major-General George O. Squier, "Developments in Radio Apparatus," Electrical World, vol. 73, 18 Jan. 1919: 129+.

[21] Yearbook for Ellington Air Base for 1918, (Houston, TX: 1918) 85-86.
"Advanced Radio Work Will Be Taught Here," Radio 8 Nov. 1918: 3.
"Pioneer Radio Technician Alfred P. Daniel Dies Here," Houston Post 1 Feb. 1955: 1.

[22] Walter H. Johnson, "Advanced Radio School," Radio 22 Nov. 1918: 4.

[23] Susan J. Douglas, Inventing American Broadcasting (Baltimore, MD: The John Hopkins University Press, 1987) 298.

[24] Lois Osborn, "Radio Broadcasting Stations of Houston, Texas," unpublished manuscript, Houston Metropolitan Research Center, 1936

[25] "City News in Brief" section, Houston Chronicle 19 Sept. 1919: 17.

[26] "Radio Operators to Send Out Air Barrage Tuesday," Houston Chronicle 28 Sept. 1919: 34.
"Who's Who in Radioland," Houston Chronicle Dec. 1925.
"City News in Brief" section, Houston Chronicle 19 Sept. 1919: 17.

[27] "Texas Coast Due for Blow Within 20 Hours says Weather Bureau," *Houston Chronicle* 14 Sept. 1919: 1.

[28] IBID.

[29] "Port Aransas Leveled, Radio Message Report," *Houston Chronicle* 15 Sept. 1919: 1.
"Storm Blows into Prominence Amateur Radio Wizard Who 'Scooped' All Weather Experts," *Houston Chronicle* 21 Sept. 1919: 11.

[30] "Texas Coast Due For Blow Within 20 Hours says Weather Bureau," *Houston Chronicle* 14 Sept. 1919: 1.

[31] Interview with Walter Peine on July 18, 1996.

[32] "Storm Blows into Prominence Amateur Radio Wizard Who 'Scooped' All Weather Experts," *Houston Chronicle* 21 Sept. 1919: 11.

[33] Susan J. Douglas, *Inventing American Broadcasting* (Baltimore, MD: The John Hopkins University Press, 1987) 298.
"Radio Operators to Send out Air Barrage Tuesday," *Houston Chronicle* 28 Sept. 1919: 34.
Texas Almanac: 2002-2003, published by *The Dallas Morning News* (Dallas, TX: distributed by Texas A & M University Press Consortium, 2002) 88.

[34] article found in *QST* August 1919.
Yearbook of Ellington Air Base for 1918 (Houston, TX: 1918) 161.

[35] "Storm Blows into Prominence Amateur Radio Wizard Who 'Scooped' All Weather Experts," *Houston Chronicle* 21 Sept. 1919: 11.

[36] Interview with Lorene Kaiser on Sept. 10, 1996.

[37] "Rice Has Complete Wireless Station," *Rice Thresher* 25 Mar. 1920: 1.

[38] United States, Department of Commerce, Bureau of Equipment, *Commercial and Government Radio Stations of the United States* (Washington, D.C.: Government Printing Office, 1920) 93.
"Radio Exchange Now a Reality," *Rice Thresher* 5 Nov. 1920: 4.

[39] "Rice Radio Club has Initial Meeting," *Rice Thresher* 24 Sept. 1920: 1.
"Officers Chosen for Radio Club," *Rice Thresher* 1 Oct. 1920: 1.
"Rice Exchange Now a Reality," *Rice Thresher* 5 Nov. 1920: 4.

[40] "Radio Exchange Now a Reality," Rice Thresher 5 Nov. 1920: 4.

[41] "Wireless Men Here Will Take Your Message," Houston Chronicle 13 Oct. 1919: 2.

Chapter 6 Endnotes

[1] "Houston Radio Club Members Can 'Talk' to Cities 1000 Miles Away," Houston Chronicle 5 Dec. 1920: 55.

[2] IBID.

[3] "Wireless Concert Fills Air; Gives Amateurs Treat," Houston Daily Post 29 Mar. 1920: 1.

[4] "Houston Radio Club Members Can 'Talk' to Cities 1000 Miles Away," Houston Chronicle 5 Dec. 1920: 55.

[5] United States, Department of Commerce, Bureau of Equipment, Amateur Radio Stations of the United States (Washington, D. C.: Government Printing Office, 1921)

[6] Interview with Walter Peine on July 18, 1996.
"Radio Experts Hear Lectures," Houston Chronicle 16 May 1920: 41.
1920 city directory for Houston.
"Houston Radio Club Members Can 'Talk' to Cities 1000 Miles Away," Houston Chronicle 5 Dec. 1920: 55.

[7] United States, Department of Commerce, Bureau of Equipment, Amateur Radio Stations of the United States (Washington, D. C.: Government Printing Office, 1921)
"Radio Experts Hear Lectures," Houston Chronicle 16 May 1920: 41.

[8] "Houston Radio Club Members Can 'Talk' to Cities 1000 Miles Away," Houston Chronicle 5 Dec. 1920: 55.
"Radio Experts Hear Lectures," Houston Chronicle 16 May 1920: 41.

[9] "Houston Radio Club Members Can 'Talk' to Cities 1000 Miles Away," Houston Chronicle 5 Dec. 1920: 55.
"Milestone in Wireless is Reached Here," Houston Chronicle 12 Feb. 1922: 50.

[10] "Who's Who in Wireless," Houston Chronicle December 1925.

[11] "Houston Radio Club Members Can 'Talk' to Cities 1000 Miles Away," Houston Chronicle 5 Dec. 1920: 55.
"Houston Radio Club Elected Officers," Houston Chronicle 17 Oct. 1920: 24.

[12] Katherine Kelley Dittmar, I Am the Family Face...An Introduction to the Autry Family & the James L. Autry Collection (New Haven, CT: K. K. Dittmar, 1992) 44.

[13] Taped interview with Harvey Wheeler on November 25, 1996

[14] full-page section in Houston Chronicle on December 14, 1919

[15] "Chronicle and Radio Club in Wireless Plan," Houston Chronicle 2 May 1920: 46.

[16] "Houston Radio Club Members Can 'Talk' to Cities 1000 Miles Away," Houston Chronicle 5 Dec. 1920: 55.

[17] "Chronicle and Radio Club in Wireless Plan," Houston Chronicle 2 May 1920: 46.

[18] "Wireless Returns on Football Game Quickens Post Service," Houston Post 25 Nov. 1921: 1.

[19] Richard Schroeder, Texas Signs On: The Early Days of Radio & Television (College Station, TX: Texas A & M University Press, 1998) 20.

[20] "Radio," The New Handbook of Texas, volume 5, ed. Roy Tyler (Austin, TX: State Historical Association, 1996) 405-06.

[21] "Hear Houston Radio Expert Across Ocean?" Houston Press 3 Nov. 1922: 8.

[22] "Wireless Telephone Concerts Will Be One of Houston Exposition Features," Houston Chronicle 16 Oct. 1921: 29.

[23] IBID.

[24] "Announce Program for Radio Concert," Houston Chronicle 4 Nov. 1922: 21.

[25] 1922 city directory for Houston

[26] "He Talks to England: Houstonian Wins Highest Radio Honors," Houston Post 4 Jan. 1923: 1
Interview with Earl Lester, Jr. on March 18, 1997.

[27] Taped interview with Roy Lay on January 29, 1997.

[28] Richard Schroeder, Texas Signs On: The Early Days of Radio & Television (College Station, TX: Texas A&M Press, 1998) 9.
Archer Gleason, History of Radio to 1926 (American Historical Society, 1938) 397.

[29] Taped interview with Roy Lay on January 20, 1997.
"Amateur Radio Student Offers Aid to Police," Houston Post 30 Jan. 1922: 24.

[30] "Post's Daily Radio Concerts Win Praise from Many Stations," Houston Post 21 May 1922: 1.

[31] Taped interview with Harvey Wheeler on November 25, 1996.
Interview with Mrs. Jerry T. Millar on November 10, 2002.

[32] David Westheimer, "First Performer Tells of Initial Broadcast," Houston Post 21 Feb. 1955: 4, section 4.

[33] Lois Osborn, "Radio Broadcasting Stations of Houston, Texas," unpublished manuscript, Houston Metropolitan Research Center, 1936, 24.
"Pioneer Radio Technician Alfred P. Daniel Dies Here," Houston Post 1 Feb. 1955: 1.

[34] "Pioneer Radio Technician Alfred P. Daniel Dies Here," Houston Post 1 Feb. 1955: 1.

[35] "Radio Club Gives Wireless Concert," Houston Chronicle 2 Jan. 1922: 9.
"To Entertain with Wireless Concert," Houston Chronicle 1 Jan. 1922: 10.

[36] "Radio Club Gives Wireless Concert," Houston Chronicle 2 Jan. 1922: 9.

[37] Lois Osborn, "Radio Broadcasting Stations of Houston, TX," unpublished manuscript, Houston Metropolitan Research Center, 1936: 25.

[38] IBID.

[39] Charlie Evans, "Greasy Garage was First Radio Station," Houston Chronicle 14 Dec. 1951: 17(B).

[40] Lois Osborn, "Radio Broadcasting Stations of Houston, TX," unpublished manuscript, Houston Metropolitan Research Center, 1936, 25.

[41] Charlie Evans, "Greasy Garage was First Radio Station," Houston Chronicle 14 Dec. 1951: 17(B).

Chapter 7 Endnotes

[1] "Houston Firm is Granted Official Radio License," Houston Post 18 Apr. 1922: 4.

[2] Erik Barnouw, A Tower in Babel: A History of Broadcasting in the United up to 1933 1st volume (New York: Oxford University Press, 1966-70) 100.

[3] "Hurlburt-Still Will Broadcast Weather Reports," Houston Chronicle 16 Apr. 1922: 17.
"Crop Reports to be Broadcasted by Local Radio," Houston Press 17 May 1922: 2.
"Houston Firm is Granted Official Radio License," Houston Post 18 Apr. 1922: 4.

[4] Interview with Walter Peine on July 18, 1996.

[5] article in Houston Post dated April 5, 1945 found in the Houston Metropolitan Research Center.

[6] Diana J. Kleiner, "Houston Post," The New Handbook of Texas, ed. Roy Tyler, vol. 3, (Austin, TX: State Historical Association) 737.
The Houston Post Story (Houston, TX.: Houston Post, 1966)

[7] Ed Kilman, "Post Extra- The Colorful History of the Houston Post" Houston Post 25 Aug. 1940.
"Post Arranges to Send Daily Radio Program," Houston Post 18 May 1922: 1.

[8] Marguerite Johnston, Houston: The Unknown City, 1836-1946 (College Station, TX: Texas A & M University Press, 1991) 218.
"Post Arranges to Send Daily Radio Programs," Houston Post 18 May 1922: 1.

[9] "Jazz and Classical Music Put on Ether by Post Broadcast," Houston Post 20 May 1922: 1.
"Post's Daily Radio Concerts Win Praise from Many Stations," Houston Post 21 May 1922: 1.

[10] "Plans Gospel Radio: Sermon to be Broadcast by Post Wireless from WCAK," Houston Post 28 May 1922: 1

[11] "Radio Artists Lengthen Concert to Furnish Jazz for South End Party," Houston Post 23 May 1922: 1.
"Arabia Temple Putting in Radio Receiving Set," Houston Post 17 May 1922: 3.

[12] "Levy Bros. to Install Radio for Customers," Houston Press 1 June 1922: 11.
"Store to Have Wireless Set," Houston Chronicle 23 Apr. 1922: 41.
"Death Takes B.J. Still, Houston Radio Pioneer," Houston Chronicle 27 Nov. 1945: 1.
"Radio Station is Being Installed at Zindler Store," Houston Post 23 Apr. 1922: 6.

[13] Interview with Thad Putnam on Sept. 20, 1996.

[14] "Iris Theater Installs Big Radio Outfit," Houston Post 14 May 1922: 8.

[15] "Will Horwitz Funeral Rites to be Held at 2:30 P.M. Today," Houston Post 26 Dec. 1941: 1.

[16] IBID.

[17] on-going advertisements, "Coolest at the Iris" in Houston Post during 1922

[18] "Radio-Equipped Car is Latest in Autos," Houston Post 2 June 1922: 42.

[19] Howard R. Smith, "Sparks' Goes to the Top," Private papers of Frank Austin III.

[20] IBID.
Interview with Maureen Wickstrom on Dec. 10, 1998.

[21] Interview with Maureen Wickstrom on Dec. 10, 1998.

[22] Howard R. Smith, "Sparks' Goes to the Top," Private Collection of Frank Austin III.

[23] "Iris Will Broadcast Lecture on Radio," Houston Post 20 May 1922: 1.

[24] "Radio Film to be Shown at Iris Tues. and Wed.," Houston Chronicle 11 June 1922: 22.
"Radio Film Open to all Houston," Houston Press 19 May 1922: 1.

[25] "WEAY Is Iris-Post Signal," Houston Post 10 June 1922: 1.

[26] IBID.

[27] "A Chronology of AM Radio Broadcasting Part 1: 1900-22," Jeff Miller's History of Radio Broadcasting, ed. Jeff Miller, <http://members.aol.com/jeff1070/chrono1.html>

[28] "Big Broadcasting Plant Here," Houston Post 14 Jan. 1923: 9.
Interview with Frank Austin III on June 12, 1997.

[29] "WEAY Is Iris-Post Signal," Houston Post 10 June 1922: 1.
"Daughters of Spain Sings for Members of Post Radio Family," Houston Post 19 June 1922: 15.

[30] "Ideal Weather for Post Radio Music Pleases," Houston Post 2 July 1922: 1.

[31] broadcasting schedule in Houston Post 5 Oct. 1922.
"Death Takes B.J. Still, Houston Radio Pioneer," Houston Chronicle 27 Nov. 1945: 1.

[32] "WEAY Is Iris-Post Signal," Houston Post 10 June 1922: 1.

[33] various broadcasting schedules in Houston Post during 1922.

[34] Taped interview with George Lacy Jr on September 7, 2002

[35] "Wireless by Kite," Southwestern Electrician Mar. 1907: 23.

[36] "Radio Reports Received at Police Bureau," Houston Chronicle 2 Mar. 1922: 6.

[37] "Fingerprint Expert to Open Radio Store," Houston Press 17 May 1922: 9.
advertisement of QRV Radio Company found in Houston Chronicle 5 Oct. 1922.

[38] "Lacy Gives Lecture on Radio Operation," Houston Press 23 June 1922: 7.
"YWCA Radio Party Proves Big Success," Houston Press 30 June 1922: 10.
"Free Lecture and Wireless Concert," Houston Press 22 June 1922: 1.

[39] "QRV Station About Ready," Houston Evening Post 30 June 1922: 3.

[40] "Radio Station WGAB Reaches Distant Points," Houston Post 6 Aug. 1922: 5.
"QRV Station About Ready," Houston Evening Post 30 June 1922: 3.

[41] various broadcasting schedules in Houston Post from July to October 1922.

[42] Taped interview with George Lacy, Jr on September 7, 2002.

[43] "Radio Station WGAB Reaches Distant Points," Houston Post 6 Aug. 1922: 5.

[44] broadcasting schedule in Houston Post on October 21, 1922.
various broadcasting schedules in Houston Post during 1922.

[45] article in Radio News August 1922: 361
United States, Department of Commerce, Bureau of Equipment, Radio Stations of the United States (Washington, D.C.: Government Printing Office, 1925) 68.

[46] "Central Student Has Broadcaster," 15 Feb. 1923 Houston Evening Post: 6.

[47] Erik Barnouv, A Tower of Babel: A History of Radio Broadcasting in the United States up to 1933 1st volume (New York: Oxford University Press, 1966-70) 125.

Chapter 8 Endnotes

[1] Erik Barnouv, A Tower of Babel: A History of Broadcasting in the United States up to 1933 1st volume (New York: Oxford Univ. Press, 1966-70) 126.

[2] "Begins Busiest Season," Houston Post 10 Sept. 1922: 40.
Marguerite Johnston, Houston: The Unknown City, 1836-1946 (College Station, TX: Texas A&M University Press, 1991) 295, 421
Mrs. W.M. Barnes, "Mrs. John Wesley Graham, Jr." Houston's Part in the World War, p. 163+

[3] Marguerite Johnston, Houston: The Unknown City, 1836-1946 (College Station, TX: Texas A&M University Press, 1991) 421.

[4] "Announce Program for Radio Concert," Houston Post 4 Nov. 1922: 21.

[5] "Post's Daily Radio Concerts Win Praise from Many Stations," Houston Post 21 May 1922: 1.
"Jazz and Classical Music Put on Ether by Post Broadcast," Houston Post 20 May 1922: 1.

[6] "Wilson Fraser," Houston Evening Post 5 July 1922.
advertisement for Texas College of Music and Art in Houston Post Sept. 10, 1922

[7] "A Brief History of the Houston Conservatory of Music," essay in author's collection.

[8] Theodore Albrecht, "Alessandro, Victor Nicholas, Jr." The New Handbook of Texas, vol. 1, ed. Roy Tyler (Austin, TX: State Historical Association) 99.

[9] "Ideal Weather for Post Radio Music Pleases," Houston Post 3 July 1922: 1.

[10] "Tom E. Stockbridge," Houston Evening Post 28 Feb. 1923.

[11] IBID.
various broadcasting schedules for WEAY in Houston Post from July 1922 to December 1922.

[12] various broadcasting schedules for WEAY in Houston Post from June 1922 to October 1922.
[13] various broadcasting schedules for WEAY in Houston Post from June 1922 to early 1923.

[14] Richard Tadlock, Jazz Masters of the 20s (New York: Macmilam Press, 1965) 174.
Charles G. Davis, "Teagarden, Weldon Jackson," The New Handbook of Texas, ed. Roy Tyler, vol. 6, (Austin, TX: State Historical Association) 229-30.
Dave Oliphant, Texas Jazz (Austin, TX: University of Texas Press, 1996) 139.

[15] "Jazz and Classical Music Put on Ether by Post Broadcast," Houston Post 20 May 1922: 1.
Richard Tadlock, Jazz Masters of the 20s (New York: Macmilam Press, 1965) 175.

[16] Dave Oliphant, Texas Jazz (Austin, TX: University of Texas Press, 1996) 139.

[17] IBID, 141.

[18] various broadcasting schedules in Houston Post during 1923.

[19] "Radio Program Gets Encore for Every Offering," Houston Post 24 Jan. 1923: 7.

[20] David Westheimer, "First Performer Tells of Initial Broadcast," Houston Post 21 Feb. 1955: section 4, page 4.

[21] "Greasy Garage was First Radio Station," Houston Chronicle 14 Dec. 1951: 17(B).

[22] "Radio Club Gives Wireless Concert," Houston Chronicle 2 Jan. 1922: 9.
"Houston Radio Fans Will Hear Complete Play," Houston Post 26 Jan. 1923: 1.

[23] "Strike News by Radio Desired by Bellville," Houston Post 2 July 1922: 3.

[24] "Pacific Coast Welterweights in Main Event," Houston Post 12 June 1922: 6.

25 "Herman-Shade Fight to be Broadcast," Houston Press 10 July 1922: 1.
26 "Thirty Fast Rounds of Boxing Promised Local Fans Tonight," Houston Post 15 Sept. 1922: 11.

27 broadcasting schedule in Houston Evening Post 28 July 1922.

28 various Sunday broadcasting schedules in Houston Post starting on July 2 1922.

29 "Iris Will Broadcast Lecture on Radio," Houston Post 20 May 1922: 3.
"Officers Chosen for Radio Club," Rice Thresher 1 Oct. 1920: 1.

30 S.E. Frost, Jr., Ph.D., Education's Own Stations: The History of Broadcast Licenses Issued to Educational Institutions (New York: Arno Press and the New York Times, 1937) 354.

31 "Police Officer George Lacy Prepares to 'Fly' After Fugitives," Houston Post 15 Oct. 1922: 8.
Taped interview with George Lacy, Jr. on September 7, 2002.

32 broadcasting schedule in Houston Post on October 21, 1922.

33 "QRV Radio Company to Install Set at Conservatory," Houston Post 24 Oct. 1922: 7.

34 "New Conservatory Radio Station to be Opened Tonight," Houston Chronicle 9 Dec. 1922: 9.
"New Radio Station to Present Program Saturday," Houston Chronicle 15 Dec. 1922: 11.

35 "Radio Concert Rice Station Monday Night," Rice Thresher 23 Feb. 1923: 1.
"Rice Men Will Bring College Atmosphere to Texas Radio Fans," Houston Chronicle 25 Feb. 1923: 55.

36 "The Second Rice Radio Concert is Successful," Rice Thresher 9 Mar. 1923: 1.
"Lecture to be Broadcasted by Radio," Rice Thresher 9 Mar. 1923: 1.
"Music Making Studies Radio Nice Program," Rice Thresher Mar. 2 1923: 1.

37 "Mammoth Program to Close Radio Year," Rice Thresher May 1923: 1.

Chapter 9 Endnotes

[1] "Big Broadcasting Plant Here: Iris-Post Station Installing New Powerful Apparatus," Houston Post 14 Jan. 1923: 9.

[2] Howard R. Smith, "Sparks Goes to the Top," Private collection of Frank Austin III.

[3] "Big Broadcasting Plant Here: Iris-Post Station Installing New Powerful Apparatus," Houston Post 14 Jan. 1923: 9.

[4] Erik Barnouv, A Tower in Babel: The History of Broadcasting in the United States up to 1933 1st volume (New York: Oxford University Press, 1966-70) 100.

[5] IBID.

[6] "Owl-Aggie Clash at Auditorium," Houston Press 16 Feb. 1923: 16.

[7] "Horwitz Starts WFO Commercial Wireless Plant," Houston Post 2 Sept. 1923: 28.

[8] Interview with Howard Van Eaton III on June 5, 1997.
1923 city directory of Houston.

[9] "Mammoth Program to Close Radio Year," Rice Thresher 26 April 1923: 1.

[10] "Rice to Radio All Games on Owl Gridiron," Rice Thresher 19 Oct. 1923: 6.

[11] S.E. Frost, Jr., Ph.D. Education's Own Stations: The History of Broadcast Licenses Issued to Educational Institutions (New York: Arno Press and the New York Times, 1937) 354.

[12] Mary Frazier, "A Life in Radio: By a Gypsy Fortune Teller," Houston Press 13 Jan. 1948: 5.

[13] Marguerite Johnston, Houston: The Unknown City, 1836-1946 (College Station, TX: Texas A & M University Press) 125.

[14] The Houston Post Story, (Houston Post: Houston, 1966) 5.

Diana J. Kleiner, "Houston Post," The New Handbook of Texas, Vol. 3, (Austin: TX: State Historical Association) 737.

[15] Taped interview with Harvey Wheeler on November 25, 1996.

[16] Death record of Ross Sterling, Jr. at Glenwood Cemetery records, Houston, Texas.

Taped interview with Harvey Wheeler on November 25, 1996.

[17] "Townsend Tells Story by Radio," Houston Post 20 July 1924: 11.

[18] promotional card of WCAK in author's possession.

[19] broadcasting schedule in Houston Post on June 26, 1924.

promotional card of WCAK in author's possession.

[20] "United States Pioneer Broadcast Service Stations," Early Radio History, ed. Thomas H. White, <http//earlyradiohistory.us/pion622l.htm>

[21] Howard R. Smith, "Sparks Goes to the Top," Private collection of Frank Austin III.

copies of radio licenses of Francis Austin II in private collection of Frank Austin III

"Horwitz Starts WFO Commercial Wireless Plant," Houston Post 2 Sept. 1923: 28.

[22] "Houston Gets Big Wireless," Houston Post 2 June 1924.

"WEAY Has New Equipment," Houston Press 11 Sept. 1924: 2.

[23] "Radiocast Fight: WEAY to Have Direct Wire for Battle," Houston Press 11 Sept. 1924: 1.

advertisement to announce hearing the Sept. 11 Firpo-Wills fight in the city auditorium printed in Houston Press on Sept. 10, 1924 on page 8

[24] Bascom Timmons, Jesse Jones: The Man and Statesman (Houston, TX: Owl Publications, 1956) 79-80.

[25] Taped interview with Harvey Wheeler on November 25, 1996.

[26] IBID.

[27] various broadcasting schedules in Houston Press during 1924.

[28] various Sunday broadcasting schedules in Houston Post starting in June, 1924.

[29] Norman W. Spellman, "Smith, Angie Frank," The New Handbook of Texas, ed. Roy Tyler, vol. 5, (Austin, TX: State Historical Association) 1089.
various Sunday broadcasting schedules in Houston Post starting in June, 1924.

[30] various Sunday broadcasting schedules in Houston Post starting in June, 1924.

[31] "Air Treats for Radio Fans," Houston Press 31 Oct. 1924: 9.

[32] "Don't Miss the Press Football Extra Tonight Rice-Texas U Game Play by Play," Houston Press 1 Nov. 1924: 1

[33] "Radiocast Game: WEAY Tells World of Texas Defeat," Rice Thresher 7 Nov. 1924: 1.

[34] "WEAY to Broadcast Sermon of Dr. Jones," Houston Post-Dispatch 4 Apr. 1925: 9.

[35] copies of Francis Austin II's radio licenses in private collection of Frank Austin III.

Chapter 10 Endnotes

[1] various broadcasting schedules in Houston Press from 1924 to 1925.

[2] Marguerite Johnston, Houston: The Unknown City, 1836-1946 1st edition (College Station, TX: Texas A & M University Press, 1991) 241.

[3] Mary Frazier, "A Life in Radio: By a Gypsy Fortune Teller," Houston Press 13 Jan. 1948: 5.

[4] IBID.
article in Houston Post dated 5 Apr. 1945 found in Houston Metropolitan Research Center.

[5] Richard Schroeder, Texas Signs On: The Early Days of Radio and Television (College Station, TX: Texas A&M University Press, 1998) 16-18, 27.
Taped interview with Harvey Wheeler on November 25, 1996.

[6] "Post-Dispatch Radio Station Will Broadcast First Program Saturday," Houston Post-Dispatch 9 May 1925: 1.

[7] compiled by the Writers' Program of the Workers Projects Administration, Houston: A History & Guide, 1942, page 210.
Marguerite Johnston, Houston: The Unknown City, 1836-1946 1st edition (College Station, TX: Texas A & M University Press, 1991)

[8] Taped interview with Harvey Wheeler on November 25, 1996.
Mary Frazier, "A Life in Radio: By a Gypsy Fortune Teller," Houston Press 13 Jan. 1948: 5.

[9] "Impressive Array of Talent on Hand When Program is Broadcast," Houston Post-Dispatch 10 May 1925: 10.

[10] license for Land Radio Station KPRC, private collection of Oveta Culp Hobby papers.
article in Houston Post dated 5 April 1945 found in the Houston Metropolitan Research Center.
"Impressive Array of Talent on Hand When Program is Broadcast," Houston Post-Dispatch 10 May 1925: 10.

[11] Taped interview with Harvey Wheeler on November 25, 1996

Jack Harris, Paul Huhndorff and Jack McGrew, The Fault Does Not Lie with Your Set: The First Forty Years of Houston Television (Austin, TX: Eakin Press, 1989) 2.
various broadcasting schedules for KPRC printed in Houston Post-Dispatch after 1926.

[12] Taped interview with Harvey Wheeler on November 25, 1996.

[13] David McComb, Houston: The Magnolia City (Austin, TX: University of Texas Press, 1986) 150.

[14] IBID.

[15] "Radio Fans Hear A.A.C.W. Program," Houston Post-Dispatch 11 May 1925: 1.

[16] article in Houston Post dated 5 April 1945 found in the Houston Metropolitan Research Center.

[17] IBID.

[18] various broadcasting schedules in Houston Post-Dispatch during 1925.

[19] "St. Paul's Musicians and Jazz Composer Entertain for KPRC" Houston Post-Dispatch 28 May 1925: 7.
various broadcasting schedules in Houston-Post Dispatch during 1925.

[20] IBID.

[21] broadcasting schedule in Houston Post-Dispatch on July 9, 1925.
broadcasting schedule in Houston Post-Dispatch on July 21, 1925.
broadcasting schedule in Houston Post-Dispatch on July 22, 1925.

[22] Section of "Dial Watchem" in Houston Post 11 April 1969
various broadcasting schedules in Houston Post-Dispatch during 1925.

[23] "Lewis, Judd Mortimer," The New Handbook of Texas, ed. Roy Tyler, vol. 4, (Austin, TX: State Historical Association) 179-80.
various broadcasting schedules in Houston Post-Dispatch during 1925.

[24] various Sunday broadcasting schedules in Houston Post-Dispatch during 1925.

[25] "Psychic Solves Mystery via Radio," Houston Post-Dispatch 19 May 1925: 1.

[26] Robert Reed, Colt .45s: A Six-Gun Salute (Houston, TX: Lone Star Books, 1999) 6.

[27] "KPRC Broadcasts Houston Ball Game to Fans' Delight," Houston Post Dispatch. 20 July 1925: 3.

[28] IBID.
Bill O'Neal, The Texas League, 1888-1987: A Century of Baseball (Austin, TX: Eakin Press, 1987) 339.

[29] "Offices and Homes Become Radio Baseball Bleachers." Houston Post-Dispatch 8 Oct. 1925.
"Dixie Series Broadcast from KPRC Bring Many Compliments; World Series Starts Wednesday," 4 Oct. 1925.

[30] "Office and Homes Become Radio Baseball Bleachers," Houston Post-Dispatch 8 Oct. 1925.

[31] "Press Box," Rice Thresher 23 Oct. 1925: 3.

Chapter 11 Endnotes

[1] Lois Osburn, "Radio Broadcasting Stations of Houston, Texas," unpublished manuscript, Houston Metropolitan Research Center, 1936, 23.

[2] Butler, Gary, "Post-War Period," Texas National Guard, 1838-1939 <http://www.kwanah.com/txmilmus/tnghist22.htm>

"Cavalry Brigade Radio Station to Go on Air Today," Houston Post 18 Feb. 1926: 7.

Lois Osburn, "Radio Broadcasting Stations of Houston, Texas," unpublished manuscript, Houston Metropolitan Research Center, 1936, 23.

"Houston's Own Cavalrymen in Action," Houston Chronicle 1 June 1924: 45.

[3] "Houston's Own Cavalrymen in Action," Houston Chronicle 1 June 1924: 45.

[4] United States,.Department of Commerce, Bureau of Equipment Radio Stations of the United States (Washington, D.C.: Government Printing Office, 1925) 60.

[5] "Cavalry Brigade Radio Station to Go on Air Today," Houston Post 18 Feb. 1926: 7.

"Station KFVI to Take Air on Thursday," Houston Chronicle. 17 Feb. 1926.

[6] "Food Talks Now Offered as Part of KFVI Program," Houston Chronicle 12 Sept. 1926: 20.

"Mother's Day Program to be Given Over Station KFVI by East End Entertainers," Houston Chronicle 9 May 1926: 50.

"Cavalry Brigade Radio Station to Go on Air Today," Houston Post 18 Feb.1926: 7.

[7] "Varied Program Will Be Put On Air by Station KFVI Sunday afternoon," Houston Chronicle 21 Mar. 1926: 33.

Richard Schroeder, Texas Signs On: The Early Days of Radio & Television (College Station, TX: Texas A & M Press, 1998) 13.

[8] "Hills Named Director of New Station," Houston Post-Dispatch 2 Feb. 1930: 1.

"Pioneer Radioman T.P. Hills, 69, Dies," Houston Chronicle. 21 Dec. 1960: 19, section 1.
"Ted Hills, 69, Pioneer in Radio, Dead," Houston Post 21 Dec. 1960: 12, section 3.
Taped interview with Harvey Wheeler on November 25, 1996.

[9] Lois Osborn, "Radio Broadcasting Stations of Houston, Texas," unpublished manuscript, Houston Metropolitan Research Center, 1936, 23.
"Jacob Schwarz Dies at Home of Daughter," Houston Chronicle 28 Ap. 1942: 8.
Obituary of Jacob Schwarz in Houston Post 29 Ap. 1942.
1913 and 1915 city directories of Houston.

[10] "Aldridge Kidd to Appear on KFVI Program," Houston Chronicle 16 June 1926: 28.
"Aldridge Kidd Presented by Station KFVI," Houston Chronicle 17 June 1926: 17.
"KFVI Artists Receive Many Request Calls," Houston Chronicle 24 June 1926: 13.
"Jacob Schwarz Dies at Home of Daughter," Houston Chronicle 28 Ap. 1942: 8.

[11] "Varied Program Will Be Put on Air by Station KFVI Sunday Afternoon," Houston Chronicle 21 Mar. 1926: 33.

[12] "Kern Tips, Voice of SWC Football, Dies," Houston Chronicle 4 Aug. 1967: 1.
Sylvia Gunn, "Tips, Kern," The New Handbook of Texas, vol. 6, ed. Roy Tyler, (Austin, TX: State Historical Association) 506.
Walter Cronkite, A Reporter's Life 1st edition (New York: A.A. Knopf, 1996) 36.

[13] "Aldridge Kidd to Appear on KFVI Program," Houston Chronicle 16 June 1926: 28

[14] "New Studio of KFVI Will be Opened Tonight," Houston Chronicle 4 May 1927: 20.
1927 city directory of Houston.

[15] "Greater KTUE to Have Formal Opening Monday, December 23," Houston Chronicle 15 Dec. 1929: 2.
"New KTUE Makes Debut; Dr. Jacobs Guest Announcer," Houston Chronicle 24 Dec. 1929: 1.

[16] "Greater KTUE Have Formal Opening Monday, December 23," Houston Chronicle 15 Dec. 1929: 2.

[17] "New KTUE Makes Debut; Dr. Jacobs Guest Announcer," Houston Chronicle 24 Dec. 1929: 1

[18] broadcasting schedule in Houston Chronicle on December 28, 1929.

[19] "Radio Man Has Had Years of Training," Houston Post-Dispatch 19 Sept. 1926: 12.
"Greater KTUE to Have Formal Opening Monday, December 23," Houston Chronicle 15 Dec. 1929: 2.

[20] "Who's Who in Radioland," Houston Chronicle 29 Nov. 1925: 31.
"Greater KTUE to Have Formal Opening Monday, December 23," Houston Chronicle 15 Dec. 1929: 2.

[21] "Greater KTUE to Have Formal Opening Monday, December 23," Houston Chronicle 15 Dec. 1929: 2.
Mary Lou Widmer, New Orleans in the Twenties (Gretna, LA: Pelican Publishing Co., 1993) 89.
Interview with Joseph Uhalt on Sept. 6, 1996.

[22] "New Broadcaster in Houston to Go on Air in 10 Days," Houston Chronicle 22 Aug. 1926: 19.
"Radio Man Has Had Years of Training," Houston Post-Dispatch 19 Sept. 1926: 12.

[23] "New Broadcaster in Houston to Go on Air in 10 Days," Houston Chronicle 22 Aug. 1926: 19.
"Local Radio Firm Moves in New Store," Houston Post-Dispatch 20 Jan. 1929, Radio Section: 1.

[24] "Greater KTUE to Have Formal Opening Monday, December 23," Houston Chronicle 15 Dec. 1929: 2.
Mary Lou Widmer, New Orleans in the Twenties (Gretna, LA: Pelican Publishing Co., 1993) 89.

[25] "Noted Speakers and Famous Musicians on First Program Over Post-Dispatch Station," Houston Post-Dispatch 15 Nov. 1929, Second KPRC Section: 6.

[26] Erik Barnouv, A Tower in Babel: The History of Broadcasting in the United States up to 1933 1st volume (New York: Oxford Press, 1966-70) 146.

[27] "Kotton Port Rail Center: Synopsis of Program for Monday," Houston Post-Dispatch 3 Oct. 1927: 4.
Richard Schroeder, Texas Signs On: The Early Days of Radio & Television (College Station, TX: Texas A & M Press, 1998) 94.

[28] "KPRC Houston-Post Dispatch: Program for Saturday, December 24, 1927, Christmas Eve," Houston Post-Dispatch 24 Dec. 1927.
Richard Schroeder, Texas Signs On: The Early Days of Radio & Television (College Station, TX: Texas A & M Press, 1998) 94.

[29] Erik Barnouv, A Tower of Babel: The History of Broadcasting in the United States up to 1933 1st volume (New York: Oxford Press, 1966-70) 184-88.

[30] "KPRC to Broadcast Dodge Hour," Houston Post-Dispatch 4 Jan. 1928: 8.
Erik Barnouv, A Tower of Babel: The History of Broadcasting in the United States up to 1933 1st volume (New York: Oxford Press, 1966-70) 191.

[31] "KPRC Houston-Post Dispatch Radio Broadcasting Station Program for Thursday, January 5, 1928," Houston Post-Dispatch 5 Jan. 1928: 15.

[32] "Complete New Equipment Has Been Installed to Give KPRC More Power for Broadcasts," Houston Post-Dispatch 8 July 1928: 1.
"Noted Speakers & Famous Musicians on First Program over Post-Dispatch Station," Houston Post-Dispatch 15 Nov. 1929, Second KPRC Section: 6.

[33] "Noted Speakers & Famous Musicians on First Program over Post-Dispatch Station," Houston Post-Dispatch 15 Nov. 1929, Second KPRC Section: 6.

[34] David Westheimer, "First Performer Tells of Initial Broadcast," Houston Post 21 Feb. 1955: 4, section 4.

[35] various broadcasting schedules in Houston Post-Dispatch during 1927.

[36] "KPRC Houston Post-Dispatch Radio Broadcasting Station Program for Monday, June 25, 1928," Houston Post-Dispatch 25 June 1928: 7.

[37] various broadcasting schedules in Houston Post-Dispatch during 1928. Chester R. Burns, "Flickwir, A.H." The New Handbook of Texas, vol. 2, ed. Roy Tyler, (Austin, TX: State Historical Association) 1032.

[38] various broadcasting schedules in Houston Post-Dispatch during 1928.

[39] "Radio Chapel of KPRC Goes on Air Sunday," Houston Post-Dispatch 6 Jan. 1929: 4.

Chapter 12 Endnotes

[1] Taped interview with Harvey Wheeler on November 25, 1996.

[2] Charter papers of the Houston Broadcasting Company filed with Secretary of State in Austin.

[3] IBID.

[4] "New Station Here to Take Air Monday," Houston Chronicle 2 Feb. 1930: 8.
Taped interview with Harvey Wheeler on November 25, 1996.
1930 city directory of Houston.

[5] "New Station One of Best in Southland," Houston Post-Dispatch 2 Feb. 1930, Radio Section: 1.
"New Station Here to Take Air Monday," Houston Chronicle 2 Feb. 1930: 8.

[6] "Editors Furnish Nick Name and Slogan for KTLC," Houston Post-Dispatch 2 Feb. 1930, Radio Section: 1.

[7] "New Station Here to Take Air Monday," Houston Chronicle 1 Feb. 1930: 8.

[8] "Hills Named Director of New Station," Houston Post-Dispatch 2 Feb. 1930, Radio Section: 1.

[9] "King Amuse-Me Reigns as Pageant of Entertainment Opens New Radio Station," Houston Post-Dispatch 2 Feb. 1930, Radio Section: 1.

[10] "New Station One of Best in Southland," Houston Post-Dispatch 2 Feb. 1930, Radio Section: 1.

[11] Bascom N. Timmons, Jesse Jones: The Man and the Statesman (Houston, TX: Owl Publications, 1956) 155.
"Greater KTUE to Have Formal Opening Monday, December 23," Houston Chronicle 15 Dec. 1929: 2.

[12] Bascom N. Timmons, Jesse Jones: The Man and the Statesman (Houston, TX: Owl Publications, 1956) 155.

[13] "KTRH Celebrates 60 and Passes the Torch," Houston Post 27 Mar. 1990: D-1.

[14] 1929 city directory of Houston.

"KTRH Celebrates 60 and Passes the Torch," Houston Post 27 Mar. 1990: D-1.

"Rice Hotel's Radio Station Will Take Air Tuesday Night," Houston Chronicle 24 Mar. 1930: 2.

[15] Taped interview with Harvey Wheeler on November 25, 1996.

Richard Schroeder, Texas Signs On: The Early Days of Radio & Television. (College Station, TX: Texas A & M Press, 1998) 16.

[16] Taped interview with Harvey Wheeler on November 25, 1996.

"KTRH Technician," Houston Chronicle 24 Mar. 1930, KTRH Radio Section: 14.

[17] "KTRH Celebrates 60 and Passes the Torch," Houston Post 27 Mar. 1990: D-1.

"Rice Hotel's Radio Station Will Take Air Tuesday Night," Houston Chronicle 24 Mar. 1930: 2.

[18] IBID.

[19] Robert Slater, This...is CBS: A Chronicle of Sixty Years (Englewood Cliffs, NJ: Prentice-Hall Inc., 1988) 15.

"Rice Hotel's Radio Station Will Take Air Tuesday Night," Houston Chronicle 24 Mar. 1930: 2.

[20] "Greater KTUE to Have Formal Opening Monday, December 23," Houston Chronicle 15 Dec. 1929: 2.

[21] Charter Papers of KTUE Broadcasting Company, filed with Secretary of State for Texas in Austin.

"Building Company Granted Charter; Officers Silent," Houston Post-Dispatch 8 Oct. 1925: 7.

[22] Charter Papers of Harris County Broadcasting Company, filed with Secretary of State for Texas in Austin.

"Station KTUE Here Ceases Broadcasts," Houston Chronicle 1 Aug. 1930: 16.

[23] "Opening Program of Radio Station KXYZ is Tonight," Houston Chronicle 20 Oct. 1930: 1.

[24] IBID.
"Rice Hotel's Radio Station Will Take Air Tuesday Night," Houston Chronicle 24 Mar. 1930: 2.
[25] "Station KXYZ Gets 375 Phone Congratulations During First Broadcast," Houston Chronicle 21 Oct. 1930: 16.

[26] Eroll R. Bogy, Houston in Brief: Interesting Facts & Figures Pertaining to Houston, Texas' Largest City and Harris County (Houston, TX: 1932) 58.

[27] Charter papers of Houston Broadcasting Company, filed with Secretary of State for Texas in Austin.
Charter papers of Harris County Broadcasting Company, filed with Secretary of State for Texas in Austin.

[28] David G. McComb, Houston: A History (Austin, TX: University of Texas Press, 1981) 135-36.

[29] A. Pat Daniels, Texas Avenue at Main St: The Chronological Story of a City Block in Houston; the Most Significant Block in the History of Texas (Houston, TX: Allen Press, 1964) 50.
John Dunning, On the Air: The Encyclopedia of Old-Time Radio (New York: Oxford University Press, 1998) 705.

[30] Taped interview with Harvey Wheeler on November 25, 1996.
Compiled by Writers' Program of the Workers Projects Administration, Houston: A History & Guide 1942, 211.
"KTRH Celebrates 60 and Passes the Torch," Houston Post 27 Mar. 1990: D-1.

[31] "Death Takes B.J. Still, Houston Radio Pioneer," Houston Chronicle 27 Nov. 1945: 1.

[32] Interview with Walter Peine on July 18, 1996.

[33] IBID.

[34] Death obituary of Gerald R. Chinski, Houston Chronicle 15 Dec. 1980.
Interview with Maureen Elizabeth Wickstrom on February 9, 1998.
Howard R. Smith, "Sparks Goes to the Top," Private collection of Frank Austin III.

[35] "Depression-era Santa Claus," Houston Chronicle 25 Dec. 1989: 1E.

"Will Horwitz Funeral Rites to be Held at 2:30 P.M. Today," Houston Post 26 Dec. 1941: 1.
"A Walk Through Tunnel History," 002 Magazine vol. 3, June 2001: 17.

[36] "Will Horwitz Faces Liquor and Lottery Charges on Border," Houston Post 6 Ap. 1932: 1.
Gene Fowler and Bill Crawford, Border Radio (Austin, TX: Texas Monthly Press, 1987) 106-08.

[37] "Will Horwitz Funeral Rites to be Held at 2:30 P.M. Today," Houston Post 26 Dec. 1941: 1.

[38] Obituary of William John Uhalt found in Houston Post on May 21, 1936: 14.

[39] Sylvia Gunn, "Tips, Kern," The New Handbook of Texas, vol. 6, ed. Roy Tyler, (Austin, TX: State Historical Association) 506.

[40] "Kern Tips, Voice of SWC Football, Dies," Houston Chronicle 4 Aug. 1967: 1.

[41] "Pioneer Radioman T.P. Hills, 69, Dies," Houston Chronicle 21 Dec. 1960: 19, section 1.
"Ted Hills, 69, Pioneer in Radio, Dead," Houston Post 21 Dec. 1960: 12, section 3.

[42] Jack Harris, Paul Huhndorff, and Jack McGrew, The Fault Does Not Lie With Your Set: The First Forty Years of Houston Television (Austin, TX: Eakin Press, 1989) 7-8.

Bibliography

Books

Baines, W. M. Mrs. Houston's Part in the World War. Edited November 11, 1919 by Mrs. W. M. Baines, editor

Barnouw, Erik. A Tower in Babel: A History of Broadcasting in the United States up to 1933. 1st Volume (New York: Oxford University Press, 1966-70)

Bartlett, Donald L. and Steele, James B. Empire: The Life, Legend and Madness of Howard Hughes 1st edition (New York: Norton Press, 1979)

Bogy, R. Eroll. Houston in Brief: Interesting Facts & Figures Pertaining to Houston, Texas' Largest City and Harris County. (Houston, 1932)

Broeske, Pat H. Howard Hughes: The Untold Story. (New York: Dutton, 1996)

Carroll, B.H. Jr. Standard History of Houston, Texas from a Study of the Original Sources. Edited by B.H. Carroll, Jr. (Knoxville, Tenn.: H. W. Crew & Co., 1912)

Chariton, O. Wallace. The Great Texas Airship Mystery. (Plano, TX.: Woodware Publishing, 1991)

Compiled by the Writers' Program of the Workers Project Administration. Houston: A History & Guide. Sponsored by the Harris County Historical Society, Inc. (Houston, TX.: Anson Jones Press, 1942)

Cronkite, Walter. A Reporter's Life. 1st edition (New York: A. A. Knopf, 1996)

Daniels, A. Pat. Texas Avenue at Main St.: The Chronological Story of a City Block in Houston; the Most Significant Block in the History of Texas. (Houston, TX.: Allen Press, 1964)

De Forest, Lee. Father of Radio: The Autobiography of Lee de Forest. (Wilcox & Follett: Chicago, Il., 1950)

Dittmar, Katherine Kelley. I Am the Family Face…An Introduction to the Autry Family & James L. Autry Collection. (New Haven, Conn.: K. K. Dittmar, 1992)
Douglas, J. Susan. Inventing American Broadcasting, 1899-1922. (Baltimore, MD: John Hopkins University Press, 1987)

Dunning, John. On the Air: The Encyclopedia of Old-Time Radio. (New York: Oxford University Press, 1998)

Emmons, E. Frederick. American Passenger Ships: The Ocean Liners and Lines, 1873-1983. (Newark: University of Delware Press; London: Associated University Presses, 1985)

Fowler, Gene and Crawford, Bill. Border Radio. (Austin, TX.: Texas Monthly Press, 1987)

Frost, S. E., Jr. PhD. Education's Own Stations: The History of Broadcast Licenses Issued to Educational Institutions. (New York: Arno Press and the New York Times, 1987)

James, Marquis. The Texaco Story: The First Fifty Years, 1902-1952. (The Texas Co.: New York, 1953)

Larson, Erik. Isaac's Storm: A Man, A Time and the Deadliest Hurricane in History. (New York: Crown Publishers, 1999)

Lewis, Tom. Empire of the Air: The Men Who Made Radio. 1st edition (New York, New York: Edward Burlingame Books, 1991)

Harris, Jack; Huhndorff, Paul and Jack McGrew. The Fault Does Not Lie With Your Set: The First Forty Years of Houston Television. (Austin, TX.: Eakin Press, 1989)

Howeth, Captain L. S. USN. History of Communications-Electronics in the United States Navy. Under the auspices of the Bureau of Ships & Office of Naval History (Washington, DC: Government Printing Office, 1963)

Johnston, Maguerite. Houston: The Unknown City, 1836-1946. 1st edition _(College Station, TX.: Texas A & M University, 1991)

McComb, David. Galveston: A History. (Austin, TX.: University of Texas Press, 1986)

McComb, David. Houston: The Magnolia City. (Austin, TX.: University of Texas Press, 1986)

Oliphant, Dave. Texas Jazz. 1st edition (Austin, TX.: University of Texas Press, 1996)

O'Neal, Bill. The Texas League 1888-1987: A Century of Baseball. (Austin, TX.: Eakin Press, 1987)

Reed, Robert. Colt 45s, A Six-Gun Salute: An Illustrated History of the Colt .45s. (Houston, TX.: Gulf Publishing Co., 1999)

Schroeder, Richard. Texas Signs On: The Early Days of Radio & Television. (College Station, TX.: Texas A & M University Press, 1998)

Slater, Robert. This...Is CBS: A Chronicle of Sixty Years. (Englewood Cliffs, NJ.: Prentice-Hall, 1988)

Tadlock, Richard. Jazz Master of the 20s. (New York: Macmilan Press, 1965)

Timmons, Bascom. Jesse Jones: The Man and the Statesman. (Houston, TX.: Owl Publications, 1956)

Widmer, Mary Lou. New Orleans in the Twenties. (Gretna, LA.:Pelican Publishing Co., 1993)

Young, S. O. Dr. A Thumbnail History of the City of Houston, TX., From Its Founding in 1836 to the Year 1912. (Knoxville, Tenn.: H. W. Crew & Co., 1912)

Periodicals and Journals

Austin Statesmen newspaper holdings on microfilm at the University of Houston-Central, Anderson Library

Bryan-College Station Eagle

Flyleaf journal holdings at Rice University, Woodson Research Center

Electrical World journal holdings at the University of Houston-Central, Anderson Library

Dallas Morning News newspaper holdings on microfilm at the University of Houston-Central, Anderson Library
Galveston Daily News newspaper holdings on microfilm at Rice University, Fondren Library

Houston Chronicle newspaper holdings on microfilm at Houston Public Central Library

Houston Chronicle and Herald newspaper holdings on microfilm at Houston Public Central Library

Houston Daily Post newspaper holdings on microfilm at Houston Public Central Library

Houston Dispatch-Post newspaper holdings on microfilm at Houston Public Central Library

Houston Evening Post newspaper holdings on microfilm at Houston Public Central Library

Houston Press newspaper holdings on microfilm at Houston Public Central Library

New Handbook of Texas

QST journal holdings on microfilm at San Antonio Public Central Library

Radio journal holdings at the University of Texas-Austin, Center for American History

Radio News journal holdings at the University of Houston-Central, Anderson Library

Rice Thresher newspaper holdings at Rice University, Woodson Research Center

San Antonio Press newspaper holdings on microfilm at San Antonio Public Central Library

Southwestern Electrician journal holdings on microfilm at University of Texas-Austin, Center for American History

Success Magazine

Victoria Daily Advocate newspaper holdings on microfilm at Houston Public Central Library, Houston Metropolitan Research Center

Miscellaneous Materials and Private Collections

Annual publications of List of Wireless Telegraph Stations on the World 1907-1910, Navy Department, Department of Commerce, Bureau of Equipment (Washington, DC: Government Printing Office)

Annual publications of Radio Stations of the United States 1915-1925, Navy Department, Department of Commerce, Bureau of Equipment (Washington, DC: Government Printing Office)

Annual publications of Commercial and Government Radio Stations of the United States 1920-1925,Navy Department, Department of Commerce, Bureau of Equipment (Washington, DC: Government Printing Office)

Annual publications of Amateur Radios Stations of the United States Navy Department, Department of Commerce, Bureau of Equipment (Washington, DC: Government Printing Office)

City directories for Houston, Texas, 1906-1930, Houston Public Central Library, Houston Metropolitan Research Center

Unpublished manuscript by Frank Matejka, The First Football Game Broadcast: Texas A & M vs. Texas University, November 24, 1921 Texas A & M University Library

Unpublished manuscript by Lois Osborn, Radio Broadcasting Stations of Houston, Texas Houston Public Central Library, Houston Metropolitan Research Center

1918 Yearbook for Ellington Air Base (Houston, TX: 1918) Houston Public Central Library, Houston Metropolitan Research Center

Interviews

All interviews mentioned as sources in endnotes were conducted by the author.

Index

5

A

B

E

F

G

H

I

T

U

V

W

X

Y